国外油气勘探开发新进展丛书（十二）

水力压裂化学剂与液体技术

［美］Johannes Karl Fink 著

卢拥军 陈彦东 等译

石油工业出版社

内 容 提 要

本书介绍了用于压裂液的各种类型液体和技术，同时还论述了自生热系统、可交联合成聚合物、单相微乳液和复合交联剂等特殊类型压裂液添加剂，在很大程度上反映了压裂液化学的最新进展和成果。

本书适合从事压裂液及其添加剂研发的专业技术人员、现场应用工程师及相关专业院校师生参考。

图书在版编目(CIP)数据

水力压裂化学剂与液体技术/(美)芬克(Fink,J.K.)著;卢拥军等译.
北京：石油工业出版社,2015.12
(国外油气勘探开发新进展丛书;12)
书名原文:Hydraulic Fracturing Chemicals and Fluids Technology
ISBN 978 – 7 – 5183 – 0958 – 0

Ⅰ.水…

Ⅱ.①芬…②卢…

Ⅲ.油层水力压裂 – 化学处理 – 处理剂 – 研究

Ⅳ.TE357.1

中国版本图书馆 CIP 数据核字(2015)第 277756 号

Hydraulic Fracturing Chemicals and Fluids Technology
Johannes Karl Fink
ISBN：978 – 0 – 12 – 411491 – 3
Copyright © 2013 by Elsevier. All rights reserved.
Authorized Simplified Chinese translation edition published by the Proprietor.
Copyright © 2015 by Elsevier (Singapore) Pte Ltd.
All rights reserved.
Published in China by Petroleum Industry Press under special arrangement with Elsevier (Singapore) Pte Ltd.. This edition is authorized for sale in China only, excluding Hong Kong SAR, Macau and Taiwan. Unauthorized export of this edition is a violation of the Copyright Act. Violation of this Law is subject to Civil and Criminal Penalties.

本书简体中文版由 Elsevier(Singapore) Pte Ltd. 授予石油工业出版社有限公司在中国大陆地区(不包括香港、澳门特别行政区以及台湾地区)出版与发行。未经许可之出口，视为违反著作权法，将受法律之制裁。

本书封底贴有 Elsevier 防伪标签，无标签者不得销售。

北京市版权局著作权合同登记号:01 – 2015 – 6806

出版发行：石油工业出版社
　　　　　(北京安定门外安华里2区1号楼　100011)
　　　　网　　址：www.petropub.com
　　　　编辑部：(010)64523710　图书营销中心：(010)64523633
经　销：全国新华书店
印　刷：北京中石油彩色印刷有限责任公司

2015年12月第1版　2015年12月第1次印刷
787×1092毫米　开本：1/16　印张：11.75
字数：300千字

定价：66.00元
(如发现装质量问题，我社图书营销中心负责调换)
版权所有，翻印必究

《国外油气勘探开发新进展丛书(十二)》
编 委 会

主 任：赵政璋

副主任：赵文智　张卫国　石　林

编　委：(按姓氏笔画排序)

　　　　王屿涛　王俊亮　卢拥军　田　冷

　　　　刘德来　李文魁　周英操　周家尧

　　　　章卫兵

序

为了及时学习国外油气勘探开发新理论、新技术和新工艺,推动中国石油上游业务技术进步,本着先进、实用、有效的原则,中国石油勘探与生产分公司和石油工业出版社组织多方力量,对国外著名出版社和知名学者最新出版的、代表最先进理论和技术水平的著作进行了引进,并翻译和出版。

从 2001 年起,在跟踪国外油气勘探、开发最新理论新技术发展和最新出版动态基础上,从生产需求出发,通过优中选优已经翻译出版了 11 辑 60 多本专著。在这套系列丛书中,有些代表了某一专业的最先进理论和技术水平,有些非常具有实用性,也是生产中所亟需。这些译著发行后,得到了企业和科研院校广大科研管理人员和师生的欢迎,并在实用中发挥了重要作用,达到了促进生产、更新知识、提高业务水平的目的。部分石油单位统一购买并配发到了相关的技术人员手中。同时中国石油天然气集团公司也筛选了部分适合基层员工学习参考的图书,列入"千万图书下基层,百万员工品书香"书目,配发到中国石油所属的 4 万余个基层队站。该套系列丛书也获得了我国出版界的认可,三次获得了中国出版工作者协会的"引进版科技类优秀图书奖",形成了规模品牌,产生了很好的社会效益。

2015 年,在前 11 辑出版的基础上,经过多次调研、筛选,又推选出了国外最新出版的 6 本专著,即《采油采气工程指南》《阿萨巴斯卡油砂矿中沥青开采原理及实践(卷一:理论基础)》《稠油及油砂提高采收率方法》《煤层气——能源与环境》《控压钻井——建模、模拟与设计》《水力压裂化学剂与液体技术》,以飨读者。

在本套丛书的引进、翻译和出版过程中,中国石油勘探与生产分公司和石油工业出版社组织了一批著名专家、教授和有丰富实践经验的工程技术人员担任翻译和审校人员,使得该套丛书能以较高的质量和效率翻译出版,并和广大读者见面。

希望该套丛书在相关企业、科研单位、院校的生产和科研中发挥应有的作用。

中国石油天然气集团公司副总经理

译者前言

随着低渗透致密砂岩油气藏和页岩油气等非常规油气资源成为人们关注的热点和投资重点，压裂改造也成为油气资源开发领域中一项重要且必不可少的工艺技术。压裂液是直接影响压裂成败的关键因素之一，压裂液化学更是涉及力学、化学、生物学、环境科学等多个学科领域。

本书是Johannes Karl Fink于2013年编撰的压裂液化学及技术的专业书籍，在很大程度上反映了近几十年来压裂液化学在理论、技术和现场应用取得的最新进展和成果，是从事压裂液及其添加剂研发的专业技术人员及现场应用的工程师的重要参考书。

本书从压裂过程中面临的各种难题出发，系统探讨了压裂液化学的最新进展。总结了用于压裂的各种类型液体和技术，并按照添加剂类别分别讨论了稠化剂、降阻剂、滤失控制剂、乳化剂、破乳剂、黏土稳定剂、pH值控制剂、表面活性剂、阻垢剂、发泡剂、消泡剂、交联剂、凝胶稳定剂、破胶剂、杀菌剂以及支撑剂等全部压裂液添加剂的性能、类别、化学组成以及主要生产商的商品名称、应用范围等；并且讨论了页岩油气藏、煤层气等特殊油藏对压裂液性能的要求及近期研究进展。同时本书还论述了自生热系统、可交联合成聚合物、单相微乳液和复合交联剂等特殊类型压裂液添加剂，并且从风险分析入手，谨慎论述了污水回收、绿色配方和自降解泡沫成分等改善压裂过程中环境问题的解决方案。书中强调了在压裂液化学和应用技术的研究方法上需要考虑多种因素综合影响，需要采用多学科交叉渗透，同时还强调了压裂液化学的研究必须进行工艺技术及油气藏地质和完井工程等的组合研究，揭示了压裂液研究向环境友好的发展趋势，压裂液的研究成为压裂技术进步的关键因素。

全书共分为20章，全部数据及实例均来自公开发表的文献，同时每章最后附有相关文献，并给出了文献所引用的各公司的商品名称及化学组成。全书由卢拥军翻译第1、第8、第14、第20章；陈彦东翻译前言、第2、第5、第8、第11、第18、第19章和附录；王丽伟翻译第3、第4、第7、第9、第13、第14章；崔伟香翻译第6、第10、第12、第15、第16、第17章。全书由陈彦东和卢拥军审核校对，舒玉华、邱晓慧、黄立宁参与部分图表和文字的校核工作。

本书的出版得到了中国石油油气藏改造重点实验室、"国家煤层气973项目"（2009CB219607）和致密气863项目（2013AA064801）的资助。中国石油勘探与生产分公司吴奇、张守良给予了大力支持，中国石油勘探开发研究院雷群、邹才能、刘合、丁云宏等给予了指导，在此一并致谢。

由于本书译者水平有限，加之时间较紧，疏漏和不足之处在所难免，敬请读者批评和指正。

前　言

本书主要介绍了压裂液化学的最新进展。在简短介绍了压裂的基本问题后,本书主要研究了压裂液的有机化学问题。

本书解释了各个添加剂的性质及其起作用机理。这里引用的数据及实例全部来源于包括专利在内的公开发表的文献。此外,随着环境保护得到越来越多的重视,环保问题也谨慎涉及。

致　　谢

本书的编撰得到了部门主管 Wolfgang Kern 教授的持续关注和高度赞赏。我很感激我们的大学图书馆 Christian Hasenhttl 博士，Johann Delanoy 博士，Franz Jurek，Margit Keshmiri，Olores Knabl，Friedrich Scheer，Christian Slamenik 和 Renate Tschabuschnig 在文献方面给予的支持。这本书不可能是凭空编撰的。

感谢米什科尔茨大学 I. Lakatos 教授，是他提供给我这个有趣的课题。

最后，还要感谢出版商的支持，特别是 Katie Hammon 提供的支持。

目 录

1 水力压裂概述 ……………………………………………………………… (1)
   1.1 应力和裂缝 …………………………………………………………… (1)
   1.2 增产方法对比 ………………………………………………………… (2)
   1.3 模拟方法 ……………………………………………………………… (3)
   1.4 测试 …………………………………………………………………… (4)
   1.5 特殊应用 ……………………………………………………………… (6)
   参考文献 …………………………………………………………………… (9)

2 液体类型 …………………………………………………………………… (12)
   2.1 不同技术的比较 ……………………………………………………… (15)
   2.2 评价专家系统 ………………………………………………………… (15)
   2.3 油基压裂液体系 ……………………………………………………… (15)
   2.4 泡沫压裂液 …………………………………………………………… (16)
   2.5 酸压裂 ………………………………………………………………… (16)
   2.6 特殊问题 ……………………………………………………………… (17)
   2.7 压裂液的表征 ………………………………………………………… (20)
   参考文献 …………………………………………………………………… (21)

3 稠化剂 ……………………………………………………………………… (25)
   3.1 水基压裂液用稠化剂 ………………………………………………… (26)
   3.2 浓缩液 ………………………………………………………………… (33)
   3.3 油基压裂液用稠化剂 ………………………………………………… (34)
   3.4 黏弹性 ………………………………………………………………… (35)
   参考文献 …………………………………………………………………… (38)

4 降阻剂 ……………………………………………………………………… (42)
   4.1 不配伍性 ……………………………………………………………… (42)
   4.2 聚合物 ………………………………………………………………… (42)
   4.3 环境因素 ……………………………………………………………… (43)
   4.4 二氧化碳泡沫压裂液 ………………………………………………… (43)
   4.5 油外相共聚物乳液 …………………………………………………… (44)

4.6 带有弱不稳定链接的聚丙烯酰胺 ……………………………………………… (45)
　　参考文献 ……………………………………………………………………… (46)

5 液体滤失添加剂 …………………………………………………………………… (47)
5.1 液体滤失剂的作用机理 …………………………………………………… (47)
5.2 化学添加剂 ………………………………………………………………… (49)
　　参考文献 ……………………………………………………………………… (54)

6 乳化剂 ………………………………………………………………………………… (57)
6.1 水包油乳液 ………………………………………………………………… (57)
6.2 反向乳液 …………………………………………………………………… (57)
6.3 水—水乳液 ………………………………………………………………… (58)
6.4 油包水包油乳液 …………………………………………………………… (58)
6.5 微乳液 ……………………………………………………………………… (58)
6.6 固相稳定乳液 ……………………………………………………………… (59)
6.7 生物处理的乳液 …………………………………………………………… (60)
　　参考文献 ……………………………………………………………………… (61)

7 破乳剂 ………………………………………………………………………………… (63)
7.1 破乳剂基本内容 …………………………………………………………… (63)
7.2 化学试剂 …………………………………………………………………… (64)
7.3 螯合剂 ……………………………………………………………………… (64)
　　参考文献 ……………………………………………………………………… (65)

8 黏土稳定剂 …………………………………………………………………………… (67)
8.1 黏土的特征 ………………………………………………………………… (67)
8.2 导致不稳定的机理 ………………………………………………………… (70)
8.3 膨胀抑制剂 ………………………………………………………………… (71)
　　参考文献 ……………………………………………………………………… (77)

9 pH 值控制剂 ………………………………………………………………………… (82)
9.1 缓冲理论 …………………………………………………………………… (82)
9.2 pH 值控制 ………………………………………………………………… (85)
　　参考文献 ……………………………………………………………………… (85)

10 表面活性剂 …………………………………………………………………………… (86)
10.1 表面活性剂性能 …………………………………………………………… (86)
10.2 黏弹性表面活性剂 ………………………………………………………… (86)

| 参考文献 ·· (89) |

11　阻垢剂 ·· (91)
| 11.1　分类及机理 ·· (91) |
| 11.2　数学模型 ··· (93) |
| 11.3　抑制剂化学 ·· (94) |
| 参考文献 ·· (99) |

12　发泡剂 ·· (104)
| 12.1　环境安全型流体 ··· (105) |
| 12.2　液态二氧化碳泡沫 ·· (105) |
| 参考文献 ··· (106) |

13　消泡剂 ·· (107)
| 13.1　消泡原理 ··· (107) |
| 13.2　消泡剂的分类 ·· (108) |
| 参考文献 ··· (110) |

14　交联剂 ·· (112)
| 14.1　交联反应动力学 ··· (112) |
| 14.2　交联剂 ·· (112) |
| 参考文献 ··· (117) |

15　冻胶稳定剂 ·· (119)
| 15.1　化学物质 ··· (119) |
| 15.2　特殊问题 ··· (119) |
| 参考文献 ··· (122) |

16　破胶剂 ·· (123)
| 16.1　水基体系的破胶 ··· (123) |
| 16.2　氧化破胶剂 ·· (123) |
| 16.3　延迟释放的酸 ·· (124) |
| 16.4　酶破胶剂 ··· (125) |
| 16.5　胶囊破胶剂 ·· (125) |
| 16.6　用于瓜尔胶的破胶剂 ·· (126) |
| 16.7　黏弹性表面活性剂凝胶液体 ·· (129) |
| 16.8　颗粒剂 ·· (129) |
| 参考文献 ··· (132) |

17 杀菌剂 …………………………………………………………………… (137)
17.1 生长机制 …………………………………………………………… (137)
17.2 性能控制 …………………………………………………………… (140)
17.3 杀菌剂处理措施 …………………………………………………… (140)
17.4 特殊化学品 ………………………………………………………… (141)
参考文献 ………………………………………………………………… (143)

18 支撑剂 …………………………………………………………………… (144)
18.1 液体滤失 …………………………………………………………… (144)
18.2 示踪剂 ……………………………………………………………… (144)
18.3 支撑剂的成岩作用 ………………………………………………… (144)
18.4 支撑剂 ……………………………………………………………… (145)
参考文献 ………………………………………………………………… (150)

19 特殊添加剂 ……………………………………………………………… (153)
19.1 自生热系统 ………………………………………………………… (153)
19.2 可交联的合成聚合物 ……………………………………………… (154)
19.3 单相微乳液 ………………………………………………………… (154)
19.4 复合交联剂 ………………………………………………………… (154)
参考文献 ………………………………………………………………… (154)

20 环境因素 ………………………………………………………………… (156)
20.1 风险分析 …………………………………………………………… (156)
20.2 污水回收 …………………………………………………………… (156)
20.3 绿色配方 …………………………………………………………… (158)
20.4 自降解泡沫成分 …………………………………………………… (159)
参考文献 ………………………………………………………………… (160)

附录 ………………………………………………………………………… (161)
商品名称 ………………………………………………………………… (161)
有机化合物缩写 ………………………………………………………… (163)
化学药品 ………………………………………………………………… (163)
术语 ……………………………………………………………………… (166)

1 水力压裂概述

水力压裂是一种用于油气井增产的工艺方法。水力裂缝是一种在原储层基质上人工产生的叠加结构,其自身以外的地层物质不受裂缝的干扰。因此在此工艺过程中,储层的有效渗透率保持不变。

由于水力裂缝增加了井筒半径,增大了井筒和储层之间的接触面,从而使产量增加。

1.1 应力和裂缝

水力压裂是石油科学中较新的技术之一,其应用时间不超过 70 年(译者:1947 年第一口井实施水力压裂)。水力压裂的经典论述认为裂缝大致垂直于最小应力方向(Yew,1997)。对于大多数深部储层来说,最小应力为水平应力,因此压裂过程中会产生垂直方向的应力。

使用弹性理论的常用工具,根据垂直静应力和水平应力平衡来计算实际应力。例如,必须使用充满流体的介质所具有的多孔弹性常数和流体静压力来校正地静应力。可以用由泊松比校正的垂直应力来计算水平应力。在某些条件下,特别是在浅层中,可以形成水平应力或垂直应力。表 1.1 中总结了可能的应力模式。

表 1.1 裂缝中应力的模式

p_b	裂缝初始破裂压力
$3s_{H,\min}$	最小水平应力
$3s_{H,\max}$	最大水平应力(=最小水平应力+构造应力)
T	岩石材料的抗拉强度
p	孔隙压力

1.1.1 裂缝启动压力

对于储层应力的认识是确定单一裂缝启动压力必不可少的。可以使用 von Terzaghi 公式来估算此压力的上限(von Terzaghi,1923),其中:

$$p_b = 3s_{H,\min} - 3s_{H,\max} + T - p \tag{1.1}$$

表 1.1 解释了各变量的含义。闭合压力是指裂缝宽度变为零时的压力,通常等于最小水平应力。

1.1.2 压力递减分析

压裂过程中的压力响应,为施工成功提供了重要的信息。可以通过闭合时间来估计压裂液效率。

1.2 增产方法对比

除了水力压裂外,还有其他增产技术,例如酸化压裂或基岩增产处理技术,而且水力压裂也可用于煤层中增加煤层气的产量。压裂液通常分为水基、油基、醇基、乳液或泡沫基液。本文综述了水力压裂的基本原则以及用于为特定作业选定配方的指导原则(Ebinger 和 Hun,1989;Ely,1989;Lemanczyk,1991)。

聚合物水化、交联和降解是这些材料经历的主要过程。这些年来,技术进步主要集中在改善流变性能和热稳定性以及交联冻胶的破胶返排特性。

1.2.1 压裂液的作用

压裂液必须同时满足多项要求。必须在以下情况保持稳定:
(1)高温;
(2)高泵送速率;
(3)高剪切速率。

这些苛刻的条件可能会导致在压裂作业完成之前压裂液降解和支撑剂过早沉降。大多数商业上使用的压裂液为冻胶或泡沫水合液。

通常情况下,压裂液由聚合物稠化剂形成冻胶。在压裂作业期间,增稠或形成的冻胶有助于支撑剂在压裂液中的携带。压裂液注入地下岩层,用于以下目的(Kelly 等,2007):
(1)创建一条从井筒延伸到地层的通道;
(2)携带支撑剂进入裂缝,形成产出流体的通道。

1.2.2 压裂作业阶段

压裂作业可以分为4个阶段,包括注入预前置液、前置液、携砂液、顶替液。预前置液是用于与地层配伍的低黏度流体,可能含有降滤失剂、表面活性剂,并具有一定的矿化度,以防止伤害地层。注入前置液(一种不含支撑剂的黏性流体)后,产生裂缝。

裂缝形成后,必须注入支撑剂,使裂缝保持渗透性。当裂缝闭合时,滞留在缝中的支撑剂会为烃类(油气)从储层基质流入井筒提供较大的过流面积和较高的导流通道。支撑剂用于保持裂缝的开启。利用黏性流体来输送、悬浮并最终使支撑剂留在裂缝中。在水力压裂施工的剪切速率范围内,这些液体的流变学特征通常表现为幂律模型特性(译者:含有屈服应力的HB 三参数模型更为准确)。

在要获得的理想的裂缝中,沿着高度及长度两个方向支撑剂的铺置应该是均匀的。但是支撑剂在非牛顿流体中沉降的复杂性,导致裂缝下部支撑剂浓度较高。这经常导致裂缝上部和井筒缺乏足够的支撑剂分布。支撑剂的聚集、包覆、桥塞和嵌入都会导致支撑剂充填层导流能力的降低(Watters 等,2010)。

作业结束时,最终进入顶替阶段,应用顶替液和其他清理剂。实际的施工进度取决于所使用的特定系统。

在作业完成后,应尽快降低流体的黏度,以便支撑剂铺置在裂缝中,而压裂液能够通过裂缝快速返排出来。控制压裂液破胶降黏的时间很重要。此外,降解的聚合物产生的残留物应很少,避免影响流体通过裂缝的流动能力。

1.3 模拟方法

裂缝几何形态的预测是水力压裂技术中最困难的技术挑战之一(Zhang 等,2010)。

已经发表的油气井中水力压裂过程的离散元模型考虑了岩石的弹性特性和莫尔—库仑裂缝判据(Torres 和 Munoz – Castano,2007)。所建立的岩石模型,是由受到构造应力和压裂液流体静压影响的弹性梁相连接的一串 Voronoi 多边形构成的。流体压力按照类似液压柱的原理处理。结果表明,模拟过程与真实情况一致。

可以用有限元软件 ABAQUS 建立一个三维非线性流固耦合的有限元模型(Zhang 等,2010)。使用该模型对中国大庆油田一口水平井的分段压裂过程进行模拟,同时考虑射孔、井筒、固井套管、一个产层、两个隔层、微环形裂缝和横向裂缝。

这些实验数据用于数值计算。微环形裂缝和横向裂缝同时产生,在压裂过程的早期会产生典型的 T 形裂缝,然后微环形裂缝消失,仅剩下横向裂缝并向外延伸。

可以得到压裂过程中地层的孔隙压力分布和裂缝形态。模拟输出的井底压力的变化与实验数据一致(Zhang 等,2010)。

1.3.1 产能

已经开发出用于计算水力压裂井产能的模型,该模型考虑了由于压裂液滤失而造成的裂缝壁面伤害的影响(Friehauf 等,2010)。

用该模型的结果与前面三维模型的结果进行对比。这些模型假设围绕井的椭圆形或径向流动,渗透率沿方位角存在差异。产能计算的显著差异表明,先前有关流动的几何参数假设可能造成产能指数的高估。

1961 年 Levine 和 Prats 给出了相同的分析结果,说明即使没有裂缝伤害,裂缝的导流能力也是受到限制的。

该模型的简单和离散性使得它非常适合于使用电子表格进行计算,并链接到裂缝性能模型中。对侵入带伤害的清除取决于地层的毛细管性质和生产过程中通过受损带施加的压力(Friehauf 等,2010)。此外,当伤害区的渗透率降低超过 9% 时,侵入带的伤害更为严重。

1.3.2 裂缝扩展

通过注入黏性、不可压缩的牛顿型流体,对可渗透岩层中现有的二维裂缝的扩展进行建模(Fareo 和 Mason,2010)。特别指出的是,2007 年 Fitt 等人将关于不渗透岩石水力压裂的方法拓展应用到可渗透的岩石。

假定裂缝中的流体为层流。应用层流理论,导出一个关于裂缝半宽与流体压力和滤失速度关系的偏微分方程。解这个方程,得到滤失速度与沿裂缝的距离和时间的一个函数。考虑 Lie 点对称的线性组合,导出群不变解。此时边界值的问题已经改写为一对初始值的问题。并考虑了滤失速度与裂缝半宽度成正比的模型(Fareo 和 Mason,2010)。

根据 Barenblatt 方法(Barenblatt 等,1990)建立了带端部效应的人工裂缝模型(Mokryakov,2011)。研究了非黏性压裂液和不渗透岩石的特殊情况。假设非黏性流体的黏度为零。已经证明,对于有限应力,应力集中区长度也不能超过一定限度。求得应力集中区的极限形态,从而可以评价极限断裂韧性。有效断裂韧性趋向于极限值的 -0.5 次方。

1.3.3 支撑剂

由于支撑剂在很大程度上决定井筒的最终产能,是压裂施工中最关键的因素。通过有效的压裂设计可以得到理论上最优的有效裂缝区。形成有足够导流能力的人工裂缝区,能够加快储层的排采(Brannon 和 Starks,2010)。

观察到所有设计中有效裂缝区和360d累计预测产量之间存在很强的相关性。累计产量对裂缝导流能力的敏感性比对裂缝有效面积的敏感性小得多。采用支撑剂部分单层增产设计比使用常规支撑剂的裂缝作业更具有竞争性,即成本更低。

1.3.4 流体损失

为了完成压裂作业设计和实施分析,建立流体损失模型至关重要。但由于其复杂性,许多参数很难或不可能去评价,建立理论模型很困难。目前已经建立了经验和半经验模型。测试方法也会影响这些模型输入数据的质量。通过对比分析目前采用的一些模型,提出了比原油模型更适合现有数据的两个不同的模型(Clark,2010)。

1.3.5 泡沫压裂液

使用混合模型将内相作为一种颗粒状流体,对泡沫压裂液的流变性进行数值模拟(Sun 等,2012)。模拟结果表明,气相分布越均匀,泡沫质量越高,靠近井壁的速度梯度越高。

在紊流情况下,紊流动能和黏性随着泡沫质量的提高而增加。泡沫质量为63%时,泡沫压裂液的流变性急剧变化。

模拟结果表明,该混合模型对于泡沫质量大于63%的区域更适用和有效(Sun 等,2012)。

1.3.6 返排控制

为使压裂液有序返排,设计了一套控制与检测系统,可避免采用人工控制时常常出现的某些不利情况,如压力震荡、吐砂等(Feng 和 Fu,2012)。

该系统使用计算机程序采集压力和流量参数以及阀门状态,控制联机设备。这样,利用了最佳的压裂液返排方法。已经有文献介绍了实际效果。

1.4 测试

1.4.1 支撑剂的铺置

在水力压裂施工过程中,通过压裂液的高压作用使地层破裂而形成裂缝。将携带支撑剂的压裂液泵入压裂裂缝通道,以防止流体压力释放时裂缝闭合。

产能的改善主要取决于裂缝的支撑尺寸,而它是由支撑剂的输送和支撑剂的适当铺置来控制的。沉降和对流是支撑剂铺置的控制机理。实验研究了采用非牛顿流体时的支撑剂输送和铺置效率,并进行了数值模拟。

用小型玻璃模型模拟水力压裂裂缝和附加参数,例如携砂液注入速度、支撑剂浓度和聚合物流变学性质(Shokir 和 Al-Quraishi,2009)。小型玻璃模型可以方便地模拟水力压裂裂缝的流动模式,且成本低廉。所观察到的流动模式与那些由非常大的流动模型获得的流动模式非常相似。

当黏性动能与重力势能比升高时,沉降速度降低,支撑剂的铺置效率增大。非牛顿流动特性指数的增加会导致支撑剂铺置效率降低。

1.4.2 滑溜水压裂

通常使用低伤害压裂液来更好地控制裂缝尺寸并产生较少的残留物。这不仅使得裂缝长度更长,而且提高了裂缝导流能力。在20世纪80年代开发的滑溜水压裂工艺比冻胶压裂液的价格更低廉。

滑溜水可以减少压裂液和支撑剂用量,并能显著增大压裂液流速。与传统的冻胶压裂液相比,滑溜水压裂后的生产效果更好(Shah和Kamel,2010)。

滑溜水压裂已经越来越多地应用于非常规页岩气藏的增产(Cheng,2012)。与交联液相比,滑溜水用作压裂液具有4个优点,其中包括成本低、更可能形成复杂裂缝网络、造成的地层伤害更小并且易于清理。

在压裂施工过程中,会向地层中注入大量压裂液。即使很好回收了返排的压裂液,但仍然会有大量的水滞留在储层中。

在已经形成的人工裂缝和已连通的天然裂缝中都有水相动力,对压裂施工的效果有显著的影响。这是由相对渗透率、毛细管压力、重力差异和裂缝导流能力控制的。

已用储层模拟模型研究生产过程中裂缝含水饱和度的分布变化情况。由于毛细管压力和重力差异而导致的水渗吸作用,在含水饱和度的分布方面具有重要影响,尤其是在长期关井过程中更为明显,进而影响天然气的流动(Cheng,2012)。

1.4.3 管材腐蚀

腐蚀现象在石油行业十分普遍。在水力压裂过程中,高压管道中经常发生管材损坏(Zhang等,2012)。随着作业时间的增加,在管道内表面出现的腐蚀和腐蚀缺陷将导致严重的管材损失以及设备故障。

目前已开发了一种用于模拟压裂液导致金属材料腐蚀损耗特性的装置(Zhang等,2012)。研究了各种参数,例如多相流速度、压裂支撑剂和冲击角等引起的腐蚀破坏机理。另外,还使用微观表面测试分析了高压管道中金属材料的腐蚀破坏机理。

1.4.4 压裂液滤失

在水力压裂裂缝扩展过程中,必须考虑沿裂缝方向以及垂直裂缝进入地层的压裂液的流动以及相关的压降(Economides等,2007)。压裂液漏失是控制裂缝长度的主要因素。

为了找到一种有效描述这种现象的数学方法,可以使用薄裂纹来表示孔压在地层中传递的边界条件。这一模型已经用于通过层间缝隙向岩层中注入热水的热传导。并提供了方程的线性解,允许进行积分变换。方程的分析解包括可以用数字计算的一些积分。

这种模型允许严格跟踪产生的裂缝数量、滤失量和增加的裂缝宽度。在注入压裂液过程中,模型有利于配方优化并允许实时计算裂缝尺寸(Economides等,2007)。

1.4.5 储层伤害

Behr等在2006年提出了水力压裂裂缝闭合后形成的复杂裂缝环境以及如何将其集成到储层模拟模型中的详细说明。开发了初始化的特殊算法并对支持工具进行测试,可以用油藏模拟程序计算致密气藏地层压裂后的性能。

为了描述裂缝几何形态和性质,将裂缝中支撑剂浓度分布以及裂缝宽度变化的信息转换为裂缝内网格化的渗透率和孔隙度。

在所考虑的滤失过程中,对压裂期间的裂缝扩展进行建模。使用表示两相非混相驱替的巴克利—莱弗里特方程对压裂液在岩石基质渗透进行建模(Behr 等,2006)。

1.4.6 交联液

用 30MPa 压力下的大规模循环流动实验,评价了硼酸盐交联瓜尔胶和硼酸盐交联泡沫压裂液的流变性和对流热交换性能(Sun 等,2010)。

结果表明,硼酸盐交联瓜尔胶在高温下发生强烈的化学降解。当温度低于阈值时,交联剂几乎失效,瓜尔胶不再交联。此外,硼酸盐交联泡沫压裂液的黏度与泡沫质量的增加成正比,而与温度的升高成反比。

流动行为指数对于井壁处非牛顿流体的速度梯度的影响是巨大的,发现了不依赖温度的对流换热系数。

剪切引起的气泡尺度的微小对流变化,可显著增加泡沫压裂液传热能力。

确定了硼酸盐交联瓜尔胶和泡沫压裂液的黏性和对流换热系数之间的关系(Sun 等,2010)。

1.5 特殊应用

非常规气藏,包括致密气、页岩气、煤层气正在成为目前和未来天然气供应的一个极为重要的组成部分(Lestz 等,2007)。

然而,这些储层往往需要各自针对性的增产措施。在低渗透储层中使用水基压裂液可能会因为外来水注入地层,形成水相圈闭而导致有效裂缝半长损失。

在致密气藏中,水具有强扩散系数,水的润湿性会使该问题更为严重。在孔隙系统中,含水饱和度增大可能会制约甲烷等气态烃的流动。

在低含水饱和度的低渗透地层中,毛细管压力可能达到 10~20MPa 或更高。此外,在不饱和储层中使用水基压裂液,不断增加的储层含水饱和度也会降低渗透率。凝析油和挥发性烃类流体成分可能有助于形成水相圈闭(Lestz 等,2007)。

1.5.1 连续油管压裂

在浅井成功使用连续油管压裂,激发了人们在深井和高温井中使用连续油管进行压裂作业的探索(Cawiezel 等,2007)。

在深井中使用连续油管压裂,压裂液主要性能必须满足:在经历高剪切和高温后,摩阻压力损失低,且有足够的携带支撑剂能力。

有文献报道了中试和油田现场试验的测试结果(Cawiezel 等,2007)。这些研究促进了连续油管压裂液的优化开发。

必须控制聚合物压裂液在通过连续油管车处于弯曲和拉直状态的油管时具有较低的摩阻损失。然而现状是通过小直径油管并随后通过高剪切区域泵送压裂液时,压裂液的稳定性会显著下降。

压裂液的优选应满足这些要求:平衡获得压裂液流变稳定性的同时,最大限度减少摩阻损失。

连续油管水力喷射压裂：使用连续油管的水力喷射压裂是专门针对低渗透水平井和直井的一项独特技术。此方法使用高压压裂液形成准确定位的人工裂缝，无需使用封隔器，可节省作业时间并降低作业风险。

在文献中已经描述了水力喷射压裂工具（Justus，2007）。在压裂过程中，将压裂工具放置到需要压裂的地层中，然后通过喷嘴向地层喷射压裂液，其压力要足够高，能穿过套管和水泥环并形成一个空腔。压力必须足够高，还要能够压开空腔中有地层压力的储层。

由于进入空腔的射流必须沿着通常与进入射流方向相反的方向流出空腔，在压开地层空腔的前端会产生高的端部压力。施加于地层空腔前端的高压，导致裂缝的形成并向地层中延伸一段距离。

某些情况下，支撑剂悬浮在裂缝的压裂液中。支撑剂可能是颗粒状物质，例如砂粒、陶瓷或矾土或其他人造颗粒、核桃壳或者能悬浮于压裂液中的其他材料。支撑剂的作用是防止裂缝闭合，从而为产出流体顺利流向井筒提供输送通道。支撑剂的存在也提高了喷射压裂液对裂缝的刻蚀作用（Surjaatmadja，2010）。

有文献研究了水力喷射射孔和水力喷射造缝的机理（Tian 等，2009）。计算了在连续油管中压裂液产生的摩阻，由此可确定现场测试的压力和流速。

理论和实践结果的比较已经证明，水力参数的理论计算和现场测试数据基本一致。也已证明，连续油管设备满足现场测试的要求（Tian 等，2009）。

1.5.2 致密气

水力压裂是提高致密气藏产能的最佳技术之一（Gupta 等，2009）。对于这样的低渗透气藏，必须进行压裂液优选。

有些储层为低压储层，必须使用增能压裂液，而有些地层对水基压裂液敏感，遇水会引起黏土的膨胀和迁移。由于压裂液冻胶残渣导致的对支撑剂充填层的伤害，是压裂施工后造成低产的主要原因之一。为减少伤害并最大限度地提高增产效果，开发了新型高效压裂液。

低浓度压裂液体系是由少量的羧甲基瓜尔胶和锆交联剂组成的。交联的延迟时间可以调节，使摩阻导致的压力损失最小，以适应于深井压裂和连续油管作业。该体系可以用二氧化碳和氮气发泡增能，也可以用二元泡沫体系。该应用已有相应的应用案例介绍（Gupta 等，2009）。

二氧化碳泡沫或二氧化碳混合增能压裂液明显有利于压裂液返排和支撑剂的输送。

在压裂过程中，冷压裂液对于岩石应力减小具有较强的冷却效果，但在热应力效果、返排过程和支撑剂充填方面，应当对这些种类的压裂液进行精确评估（Rafiee 等，2009）。已经在地热储层中成功测试了冷水压裂液。在这些领域成功的增产方法并不适合于致密气藏。

使用液态二氧化碳压裂液为增产提供了一种可行的新方法。这种压裂液已经在美国的各种地层中成功应用，可作为常规增产液的经济替代品。

在某些情况下也可以使用液态氮作为压裂液。从现场应用情况看，液态二氧化碳对致密气的增产似乎是最有效的。

将冷压裂技术与地下二氧化碳捕获和储存相结合是一种令人关注的也很有前途的替代方案（Rafiee 等，2009）。

1.5.3 页岩气(油页岩与天然气水合物)

Maguire 等人在 2007 年描述了一种油页岩和天然气水合物的原位生产方法,其中将液化气体注入横向压裂井筒形成裂缝网络。随后加热,使干酪根液化或天然气水化合物离解,使得油页岩生成的石油或天然气可以通过裂缝开采。

在温度为 $-60°C$($-75°F$)条件下,以 500psi 的破裂压力注入液态氮,将使其体积增大 14 倍。如果裂缝体积不增大,在此温度下,膨胀压力将增加至约 7000psi。

由于天然气水合物的分解和水合物冰的收缩都会形成空隙,而注入水可以替代天然气水合物占据这些空隙,达到防止水合物层塌陷的目的,因此使用水作为增温剂非常重要。

注入水后,加热的水将离解天然气水合物,天然气将通过人工裂缝体系向下运移到下部生产井筒,并进入套管环空,并由此到达地表。

加热水所需的热量,可以由燃料蒸汽发生器中产出气的燃烧提供。热水由天然气水合物的顶部注入并向下运移,这样可以避免由于从下部层位注入而导致的先注入的水吸收后注入水的热量。

静水压力和增加的注入压力加上下部生产井筒压力将使释放的天然气向下流动而不是由于浮力因素向上流动(Maguire,2007)。

1.5.4 煤层气

煤层气的生产通常需要采用水力压裂增产。现有文献已介绍了对煤层各种处理方法效果方面的基础研究(Conway 和 Schraufnagel,1995;Penny 和 Conway,1995)。

在一口井中,用含有脱水剂的油井处理液对煤层进行处理。这种助剂提高了地层的渗透率,从而提高产水量,并与煤层面紧密结合,从而在较长的生产过程中实现了提高渗透率的效益。

脱水表面活性剂可以是聚氧化乙烯、聚氧化丙烯和聚碳酸亚乙酯(Nimerick 和 Hinkel,1991)或对叔戊基苯酚与甲醛缩合物,或者由 80%~100% 的烷基异丁烯酸酯单体和亲水性单体组成的共聚物(Harms 和 Scott,1993)。图 1.1 示意了用于此目的的化合物。

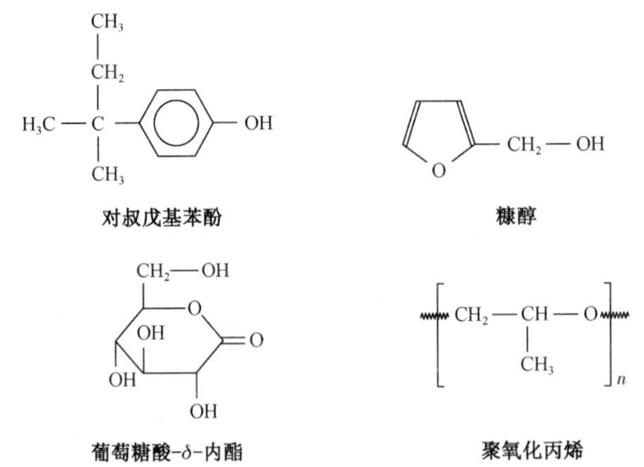

对叔戊基苯酚

糠醇

葡萄糖酸-δ-内酯

聚氧化丙烯

图 1.1 单体脱水

这样的油井处理液可以用于压裂和挑战性作业,用于长期生产过程中提高和维持裂缝导流能力。

活性水和植物胶压裂液广泛用于中国煤层气开采(Dai 等,2011)。但是,由于流变性较差、水溶性不好以及高残留,其应用受到限制。

冻胶压裂液由非离子型的聚丙烯酰胺、$ZrOCl_2$ 交联剂、pH 值调节剂、$(NH_4)_2S_2O_8$ 破胶剂以及活化剂组成。该配方用于低温(20~40°C)和低渗透煤层气储层。

这种类型的冻胶压裂液具有易于制备、低成本、强抗剪力、低滤失系数、迅速破胶、无破胶残余以及易于返排的优点。此外,冻胶压裂液的性能比活性水和植物胶更优越,非常适合于低温煤层气的压裂施工(Dai 等,2011)。

参 考 文 献

Barenblatt, G. I., Entov, V. M., Ryzhik, V. M., 1990. Theory of Fluid Flows Through Natural Rocks. Kluwer Academic Publishers, Dordrecht, Boston.

Behr, A., Mtchedlishvili, G., Friedel, T., Haefner, F., 2006. Consideration of damaged zone in atight gas reservoir model with a hydraulically fractured well. SPE Prod. Oper. 21(2), 206–211. http://dx.doi.org/10.2118/82298-PA.

Brannon, H., Starks, T. R. I. I., 2010. Less can deliver more. Oilfield Technol. 3(2), 59–63. Cawiezel, K. E., Wheeler, R. S., Vaughn, D. R., 2007. Specific fluid requirements for successfulcoiled-tubing fracturing applications. SPE Prod. Oper. 22(1), 83–93. http://dx.doi.org/10.2118/86481-PA.

Cheng, Y., 2012. Impact of water dynamics in fractures on the performance of hydraulicallyfractured wells in gas-shale reservoirs. J. Can. Petrol. Technol. 51(2), 143–151.

Clark, P. E., 2010. Analysis of fluid loss data II: models for dynamic fluid loss. J. Petrol. Sci. Eng. 70(3–4), 191–197. http://dx.doi.org/10.1016/j.petrol.2009.11.010.

Conway, M. W., Schraufnagel, R. A., 1995. The effect of fracturing fluid damage on production fromhydraulically fractured wells. In: Proceedings Volume. Alabama University et al. InternationalUnconventional Gas Symposium (Intergas 95)(Tuscaloosa, AL, 14–20 May 1995), pp. 229–236.

Dai, C., You, Q., Zhao, H., Guan, B., Wang, X., Zhao, F., 2011. A study on gel fracturing fluidfor coalbed methane at low temperatures. Energy Sources Part A: Recovery Utiliz. Environ. Effects 34(1), 82–89. http://dx.doi.org/10.1080/15567036.2010.545806.

Ebinger, C. D., Hunt, E., 1989. Keys to good fracturing: Pt. 6: new fluids help increase effectivenessof hydraulic fracturing. Oil Gas J. 87(23), 52–55. Economides, M. J., Mikhailov, D. N., Nikolaevskiy, V. N., 2007. On the problem offluid leakoff during hydraulic fracturing. Transport Porous Med. 67(3), 487–499. http://dx.doi.org/10.1007/s11242-006-9038-7.

Ely, J. W., 1989. Fracturing fluids and additives. In: Henry L. Doherty (Ed.), Recent Advances inHydraulic Fracturing, vol. 12 (SPE Monogr Ser). SPE, Richardson, Texas.

Fareo, A. G., Mason, D. P., 2010. A group invariant solution for a pre-existing fluid-drivenfracture in permeable rock. Nonlinear Anal.: Real World Appl. 12(1), 767–779. http://dx.doi.org/10.1016/j.nonrwa.2010.08.004.

Feng, Y., Fu, K., 2012. Automatic control of liquid discharging after hydraulic fracture of oil well. Adv. Mater. Res. 443–444 (Pt. 2, Manufacturing Science andMaterials Engineering), 774–778. http://dx.doi.org/10.4028/www.scientific.net/AMR.443-444.774.

Fitt, A. D., Mason, D. P., Moss, E. A., 2007. Group invariant solution for a pre-existing fluid-drivenfracturein impermeable rock. Zeitschrift für angewandteMathematik und Physik 58(6), 1049–1067. http://dx.doi.org/

10. 1007/s00033 - 007 - 7038 - 2.

Friehauf, K. E. , Suri, A. , Sharma, M. M. , 2010. A simple and accurate model for well productivityfor hydraulically fractured wells. SPE Prod. Oper. 25(4), 453 - 460.

Gupta, D. V. S. , Jackson, T. L. , Hlavinka, G. J. , Evans, J. B. , Le, H. V. , Batrashkin, A. , Shaefer, M. T. , 2009. Development and field application of a low - pH, efficient fracturing fluid for tightgas fields in the Greater Green River basin, Wyoming. SPE Prod. Oper. 24(4), 602 - 610. http://dx. doi. org/10. 2118/116191 - PA.

Harms, W. M. , Scott, E. , 1993. Method for stimulating methane production from coal seams. USPatent 5 249 627, assigned to Halliburton Co. , October 5 1993. Justus, D. M. , 2007. Hydrajet perforation and fracturing tool. US Patent 7 159 660, assigned to Halliburton Energy Services, Inc. (Duncan, OK), January 9 2007. < http://www. freepatentsonline. com/7159660. html >.

Kelly, P. A. , Gabrysch, A. D. , Horner, D. N. , 2007. Stabilizing crosslinked polymer guars andmodified guar derivatives. US Patent 7 195 065, assigned to Baker Hughes Incorporated (Houston, TX), March 27 2007. < http://www. freepatentsonline. com/7195065. html >.

Lemanczyk, Z. R. , 1991. The use of polymers in well stimulation: performance, availability andeconomics. In: Proceedings Volume. Plast Rubber Institute Use of Polymers in Drilling &Oilfield Fluids Conference(London, England, 12 September 1991).

Lestz, R. S. , Wilson, L. , Taylor, R. S. , Funkhouser, G. P. , Watkins, H. , Attaway, D. , 2007. Liquidpetroleum gas fracturing fluids for unconventional gas reservoirs. J. Can. Petrol. Technol. 46 (12), 68 - 72. http://dx. doi. org/10. 2118/07 - 12 - 03.

Levine, J. S. , Prats, M. , 1961. The calculated performance of solution - gas - drive reservoirs. Soc. Petrol. Eng. J. 1 (3), 142 - 152. http://dx. doi. org/10. 2118/1520 - G.

Maguire, J. Q. , 2007. In - situ method of producing oil shale and gas (methane) hydrates, on - shoreand off - shore. US Patent 7 198 107, assigned to Maguire and James Q. (Norman, OK), April 32007. < http://www. freepatentsonline. com/7198107. html >.

Maguire, J. Q. , 2010. In - situ method of fracturing gas shale and geothermal areas. US Patent7 784 545, August 31 2010. < http://www. freepatentsonline. com/7784545. html >.

Mokryakov, V. , 2011. Analytical solution for propagation of hydraulic fracture withbarenblatt's cohesive tip zone. Int. J. Fracture 169(2), 159 - 168. http://dx. doi. org/10. 1007/s10704 - 011 - 9591 - 0.

Nimerick, K. H. , Hinkel, J. J. , 1991. Enhanced methane production from coal seams by dewatering. EP Patent 444 760, assigned to Pumptech NV and Dowell Schlumberger SA, September 41991.

Penny, G. S. , Conway, M. W. , 1995. Coordinated studies in support of hydraulic fracturing ofcoalbed methane: Final report(July 1990 - May 1995). Gas Res Inst Rep GRI - 95/0283, Gas ResInst.

Rafiee, M. M. , Wilsnack, T. , Voigt, H. D. , Haefner, F. , 2009. Cold - frac technology in tight gasreservoirs. DGMK Tagungsbericht 1 (DGMK/OGEW - Frnhjahrstagung des FachbereichesAufsuchung und Gewinnung, 2009), 441 - 450.

Shah, S. N. , Kamel, A. H. A. , 2010. Investigation of flow behavior of slickwater in large straight andcoiled tubing. SPE Prod. Oper. 25(1), 70 - 79. http://dx. doi. org/10. 2118/118949 - PA.

Shokir, E. M. , Al - Quraishi, A. A. , 2009. Experimental and numerical investigation ofproppant placement in hydraulic fractures. Petrol. Sci. Technol. 27(15), 1690 - 1703. http://dx. doi. org/10. 1080/10916460802608768.

Sun, X. , Wang, S. , Lu, Y. , 2012. Study on the rheology of foam fracturing fluid with mixturemodel. Adv. Mater. Res. 512 - 515 (Pt. 3, Renewable and Sustainable Energy II), 1747 - 1752. http://dx. doi. org/10. 4028/www. scientific. net/AMR. 512 - 515. 1747.

Sun, X. , Wang, S. - Z. , Bai, Y. , Liang, S. - S. , 2010. Rheology and convective heat transfer propertiesof borate cross - linked nitrogen foam fracturing fluid. Heat Transfer Eng. 32 (1), 69 - 79. http://dx. doi. org/

10.1080/01457631003732979.

Surjaatmadja, J. B., 2010. Hydrajet tool for ultra high erosive environment. US Patent 7 841 396, assigned to Halliburton Energy Services Inc. (Duncan, OK), November 30 2010. < http://www.freepatentsonline.com/7841396.html >.

Tian, S., Li, G., Huang, Z., Niu, J., Xia, Q., 2009. Investigation and application formultistage hydrajet - fracturing with coiled tubing. Petrol. Sci. Technol. 27(13), 1494 - 1502. http://dx.doi.org/10.1080/10916460802637569.

Torres, G. S. A., Munoz - Castano, J. D., 2007. Simulation of the hydraulic fractureprocess in two dimensions using a discrete element method. Physical Review E: Statistical, Nonlinear Soft Matter Phys. 75 (6 - 2), 066109/1 - 066109/9. http://dx.doi.org/10.1103/PhysRevE.75.066109.

von Terzaghi, K., 1923. Die Berechnung der Durchlässigkeitsziffer des Tones aus demVerlauf der hydrodynamischen Spannungserscheinungen. Sitzungsberichte der Akademie derWissenschaften in Wien, Mathematisch - Naturwissenschaftliche Klasse, Abteilung 2a.

Watters, J. T., Ammachathram, M., Watters, L. T., 2010. Method to enhance proppant conductivityfrom hydraulically fractured wells. US Patent 7 708 069, assigned to Superior Energy Services, L. L. C. (New Orleans, LA), May 4 2010. < http://www.freepatentsonline.com/7708069.html >.

Yew, Ching H., 1997. Mechanics of Hydraulic Fracturing. Gulf Pub. Co., Houston, Tex. http://www.worldcat.org/title/mechanics - of - hydraulic - fracturing/oclc/162129743&referer = brief_results.

Zhang, G. M., Liu, H., Zhang, J., Wu, H. A., Wang, X. X., 2010. Three - dimensional finite elementsimulation and parametric study for horizontal well hydraulic fracture. J. Petrol. Sci. Eng. 72 (3 - 4), 310 - 317. http://dx.doi.org/10.1016/j.petrol.2010.03.032.

Zhang, J., Fan, J., Xie, Y., Wu, H., 2012. Research on erosion of metal materials for highpressure pipelines. Adv. Mater. Res. 482 - 484(2), 1592 - 1595. http://dx.doi.org/10.4028/www.scientific.net/AMR.482 - 484.1592.

2 液体类型

一般情况下,水力压裂施工先泵注不含支撑剂的黏性液体,即前置液,通常是在水中加入一些添加剂,以获得高的黏度,并且使液体入井的速度远大于滤失入地层的速度。在井筒附近的储层内形成高压,压开地层形成裂缝并使之扩大延伸。

压开地层后,将支撑剂,如砂子,加入液体中,组成混砂液,泵注到地层中压开的裂缝里以避免在压力去除后裂缝闭合。基液的携砂能力取决于加入到水基液中添加剂的类型(Lukocs等,2007)。

自从20世纪50年代后期,超过半数的压裂施工使用的是含有瓜尔胶或瓜尔胶衍生物的液体,诸如羟丙基瓜尔胶(HPG),羟丙基纤维素(HPC),羧甲基瓜尔胶(CMG)和羧甲基羟丙基瓜尔胶(CMHPG)。

通常使用基于硼、钛、锆或铝络合物的交联剂增加聚合物的有效相对分子质量,以更好地适用于高温井。

纤维素衍生物诸如羟乙基纤维素(HEC)、羟丙基纤维素(HPC)和羧甲基羟乙基纤维素(CMHEC)等被用作稠化剂使用,此类稠化剂用或不用交联剂都行。黄胞胶和硬葡聚糖也显示出具有优良的支撑剂悬浮能力,但是比瓜尔胶衍生物更昂贵,因此很少使用。

聚丙烯酰胺(PAM)、聚丙烯酸酯聚合物和共聚物作为稠化剂通常用于高温地层,而作为降阻剂,则使用较低的浓度,应用在所有温度范围内(Lukocs等,2007)。

可以用黏弹性表面活性剂(VES)获得无聚合物水基压裂液,这些液体通常是由适量的表面活性剂混合而成,这种表面活性剂包括阴离子型、阳离子型、非离子型和两性离子表面活性剂。其黏度是由于其组分在液体中交联成了三维结构,当表面活性剂浓度超过某一临界浓度时,黏度随之增加。此时表面活性剂分子聚集形成胶束,分子间相互作用从而形成网络结构,使溶液具有黏弹性特征。

阳离子VES通常是含有长链的季铵盐,如十六烷基三甲基溴化铵是目前为止商业化最多的表面活性剂。其他在产生黏弹性的表面活性剂溶液中常见的试剂包括盐,如氯化铵、氯化钾、氯化钠、水杨酸钠、异氰酸酯钠和非离子型有机分子,如氯仿。表面活性剂溶液的电解质含量对于控制其黏弹性特征也很重要(Lukocs等,2007)。

这种类型的阳离子表面活性剂液体在高盐水浓度会失去黏性,因此其在砾石充填或钻井液中应用受限,阴离子VES也是如此。

两性离子表面活性剂(Allan等,2008年)与有机酸、有机盐或无机盐也能表现出黏弹性性能。这种表面活性剂可以是来自某些蜡、脂肪和油的二羟基烷基甘氨酸钠、烷基两性乙酸盐,或VES丙酸、烷基甜菜碱、烷基酰胺丙基甜菜碱、烷基单(二)丙酸盐。这种表面活性剂一般带有无机水溶性盐或有机添加剂,如邻苯二甲酸、水杨酸及其盐类。

两性离子表面活性剂,特别是那些含有甜菜碱基团的,可以用于高达150℃的温度,因此特别适合在中高温地层中应用。图2.1给出了甜菜碱的分子式。与前文提到的阳离子黏弹性表面活性剂相同,阴离子表面活性剂通常无法在高盐水浓度下使用。

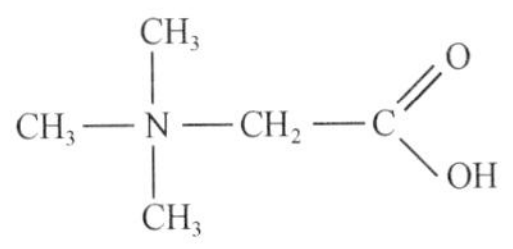

图 2.1　甜菜碱

常用无聚合物的 VES 液体减少对支撑剂带的损伤，并且将支撑剂有效地输送到裂缝中。在压裂工程中，正确评价液体的流变性能和支撑剂沉降是非常重要的（Wang 等，2012）。

已经有人研究了 VES 压裂液的流变及黏温性能。VES 液体行为特征表现为非牛顿剪切变稀流体，并且在适当的剪切速率和温度范围内可以用幂律模型描述液体的流变性能。然而，随着剪切速率和温度的增加，流体逐渐接近牛顿流体。

当 VES 浓度为 4% 时，液体会形成稳定细小的蠕虫状胶束结构，从而具有良好的黏弹性和较高的携砂能力（Wang 等，2012）。

目前已经有了表面活性基无聚合物的液体简单制备方法（Deng 等，2012）。在已测试的流体中可能的成分是十四烷基三甲基溴化铵、十六烷基三甲基溴化铵、十八烷基三甲基溴化铵和水杨酸。评价了作为压裂液相关的黏弹性和携砂能力。

测试报告指出，这些水基凝胶有很强的携砂能力和高的黏弹性能。由十八烷基三甲基溴化铵与水杨酸体系组成的凝胶体系性能最好。溴化铵与酸的比例增加，凝胶的黏弹性增加（Deng 等，2012）。

支撑剂可以是砂子、中强度陶粒支撑剂或烧结高铝矾土，也可以是涂有树脂层提高聚合能力的支撑剂。可以用树脂或如纤维类的支撑剂回流控制剂包裹。根据对支撑剂的不同性能的要求，如密度、尺寸或浓度，可以选择具有不同沉降速度的支撑剂。

清水压裂作业是用低成本、低黏度流体改造特低渗透油藏。其机理在于形成了粗糙表面（岩石剥落），岩石发生剪切位移并且局部高支撑剂浓度形成了足够的导流能力，这是措施成功的主要原因。该机理可以用类似于"楔子劈开木头"的行为来描述。

黏性措施工作液一般通过溶于水溶液的多糖或合成聚合物，与有机金属化合物形成交联体系获得。金属交联聚合物在包括水力压裂、砾石充填、堵水和其他完井作业等措施井作业中都有成功应用范例。

为了确保施工成功，在支撑剂铺置结束后，流体的黏度最终应降低至接近于水的黏度。这样可以使在开井复产后，部分压裂液返排而不会带出过多的支撑剂。如果液体的黏度低，会在地层流体的影响下自然流动。黏度的这种降低或转换被称为破胶，是通过在初始冻胶中添加称之为破胶剂的化学添加剂完成的。

有些压裂液，如那些基于瓜尔胶聚合物的，可以不用加入破胶剂而自然破胶，但是破胶时间往往要超过 24h，或者数周、数月、数年，时间长短取决于地层条件。

为了缩短破胶时间，通常在冻胶中加入化学添加剂。典型的做法是添加氧化剂或酶来降解高分子冻胶结构。诸如过硫酸盐、亚铬盐、有机过氧化物或碱土金属或过氧化锌盐等类型的氧化剂，或酶都是有效的。

破胶时间也很重要。冻胶过早破胶，悬浮的支撑剂会在液体中沉降脱离出来，这样支撑剂就无法在人工裂缝中达到足够远的距离。过早破胶还会过早降低流体的黏度,致使施工达不到适当的裂缝宽度。另外,冻胶破胶过慢会导致压裂液返排困难,恢复生产也随之延迟。

过早破胶还会出现更多其他问题,如支撑剂从裂缝中回流,至少造成部分裂缝闭合,从而降低了压裂效果。压裂冻胶最好是在泵注结束时开始破胶,并且在施工后24h内完全破胶。

压裂液的组成主要包括溶剂、水溶性或可水化的聚合物、交联剂、无机破胶剂、可选加的酯类化合物和胆碱酯。溶剂可以是氯化钾水溶液,无机破胶剂可以是金属氧化剂,如碱土金属和过渡金属,也可以是镁、钙或锌的过氧化物。酯类化合物可以是多元羧酸酯,如草酸酯、柠檬酸酯或乙二胺四乙酸酯。那些具有羟基基团还可以被乙酰化也是较好的酯类化合物,如乙酰化柠檬酸得到的乙酰柠檬酸三乙酯。

水合聚合物可以是水溶性多糖,如半乳甘露聚糖和纤维素,交联剂可以是硼酸、钛酸盐或含锆化合物,如 $Na_3BO_3 \cdot nH_2O$。

文献中给出了商业化的压裂液添加剂(Anonymous,1999)。表2.1 给出了压裂液的可能组成,说明了压裂液配方的复杂程度。某些添加剂不会一起使用,像胶凝油的添加剂就不会在水基压裂液中使用。超过90%的压裂液是水基压裂液。如果使用添加剂,水是经济的和可以在较大范围内控制物理性质的流体。压裂液添加剂有两大用途(Harris,1988):

表2.1 压裂液组成

组成/类别	性能/说明
水基聚合物	稠化剂,输砂,降低地层滤失
降阻剂	降低管路摩阻
液体滤失添加剂	如果稠化剂量不够,可以形成滤饼,降低地层滤失
破胶剂	在施工后降解稠化剂或者降低交联剂能力(不同化学机理)
乳化剂	形成柴油混合的凝胶
黏土稳定剂	用于含有黏土的地层
表面活性剂	防止水润湿地层
破乳剂	消除乳化
pH值控制剂	增加液体的稳定性(例如提高耐温性能)
交联剂	增加增稠液体的黏度
起泡剂	用于泡沫压裂液
冻胶稳定剂	保持冻胶较长的有效期
消泡剂	消除泡沫
油基胶凝剂	相当于油基压裂液的交联剂
杀菌剂	阻止生物降解
水基线性胶体系	一般体系
交联冻胶体系	增加黏度
油基体系	用于水敏地层
树脂涂层支撑剂	支撑剂材料
陶粒	支撑剂材料

(1) 增强压裂液的造缝和携砂能力；
(2) 减小地层伤害。

稠化剂（如聚合物和交联剂）、温度稳定剂、pH 值控制剂和流体滤失控制材料，协同作用压开一条裂缝。通过使用破胶剂、杀菌剂、表面活性剂、黏土稳定剂和气体可以降低地层伤害。表 2.2 总结了用于压裂的各种类型液体和技术。

表 2.2 水力压裂液的不同类型

类型	说明
水基液体	主要应用
油基液体	水敏；增加失火危险
醇基液体	很少使用
乳化液	高压，低温
泡沫液	低压，低温
简单增稠液	简化技术
氮气泡沫压裂	快速返排
复合稠化水压裂	通常为最佳解决方案
预混合浓缩胶	提高物流效率
就地沉淀技术	减少结垢问题（Hrachovy，1994a，1994b）

2.1 不同技术的比较

应用哪种压裂技术最优取决于储层类型。在相同环境中进行技术比较是可行的。在堪萨斯 Hugoton 油田（MESA 有限责任合伙公司），测试了几种水力压裂方法（Cottrell 等，1988）。

与泡沫技术和较早的简单技术相比，采用复合稠化水压裂是非常成功的一种方法，这一技术已经在 56 口井上得到了应用。

2.2 评价专家系统

基于 PC 技术，已经开发了交互式计算机模型，辅助工程师根据给定地层特征，优选压裂液、添加剂和最适合的支撑剂（Holditch 等，1993；Xiong 等，1996）。

该模型还基于储层性能和经济性优化了处理液量。为了选择压裂液、添加剂和支撑剂，专家系统咨询了来自不同公司的增产技术专家，查阅了文献，然后综合这些知识形成一定选用规则，用专家系统的外壳实现这一目标。

此外，在水力压裂过程中的液体滤失可以通过建模、计算和实验测得。文献中已经给出了把实验室数据转换为在地层条件下的滤失评估的程序（Penny 和 Conway 1989）。

2.3 油基压裂液体系

相比水基冻胶压裂液而言，胶凝油压裂液的一个优点是可以在水敏地层中使用。

2.4 泡沫压裂液

在很多压裂作业中都可以使用泡沫压裂液,尤其是在某些敏感地层中(Stacy 和 Weber,1995)。泡沫压裂液具有可以重复使用、剪切稳定、并且在很大温度范围内可以形成稳定泡沫的性能。即使在相对高温条件下,也表现出较高的黏度(Bonekamp 等,1993)。

由于具有较强的增能能力,可以提高返排量和返排速度(Tamayo et 等,2008)等优点,二氧化碳、氮气和两者混合的高质量泡沫广泛用于致密深井地层。

泡沫压裂液是相对较多的气体分散在相对较少的液体中,需要加入表面活性剂,以保证在气液混合时可以产生并获得稳定的泡沫(Welton 等,2010)。

较差的泡沫压裂液里往往有相当多的不均匀气泡尺寸分布,也就是说大小不同的气泡混合在一起。相反,较好结构的泡沫压裂液泡沫尺寸分布较为均匀,大多数气泡都相对较小(Middaugh 等,2007)。

较差的泡沫压裂液中也有较好结构的泡沫。这种压裂液在较好结构区域即使在高泡沫质量下也是可以携砂的。泡沫压裂液中一般常用的气体是氮气和二氧化碳,这是因为这两种气体是惰性的,常见且相对便宜(Welton 等,2010)。

表面活性剂是用来降低表/界面张力的,同时也是在压裂液体系设计中考虑增加压裂液返排率,减少液体在地层中滞留的关键因素(Tamayo 等,2008)。增加液体的返排率可以减少作业费用并且缩短液体返排时间,从而改善整体经济效益。最大的好处是可以减少支撑带伤害,获得较高的裂缝导流能力。

二氧化碳泡沫压裂液和表面活性剂的应用可以增加压裂液的返排率已经在现场应用中得到了证实(Tamayo 等,2008)。

还可以使用可回收泡沫压裂液(Chatterji 等,2007)。在施工完成后,压裂液 pH 值发生改变,泡沫破灭,支撑剂从液体中沉降出来。液体返排至地面后回收至罐,通过调整 pH 值至最初配制时的值,并加入气体再次形成泡沫。

2.5 酸压裂

酸压裂和基质酸化是不一样的。酸压裂适用于低渗可酸溶储层。基质酸化是用于高渗储层的技术。酸压裂的目的层一般为石灰岩($CaCO_3$)或白云岩[$CaMg(CO_3)_2$]。

这些矿物易于和盐酸反应生成氯化物和二氧化碳。相较加砂压裂技术而言,酸压裂的好处是不会出现支撑剂回流现象。酸液在裂缝表面造成的不均匀刻蚀面,可以在裂缝闭合后为地层流体流入井筒提供高的导流通道(Mukherjee 和 Cudney,1992)。

另一方面,由于酸液与地层反应很快消耗完了,酸蚀裂缝长度一般较短。如果在盐酸中含有氟化物,会生成不溶性氟化钙沉淀,从而造成地层堵塞,影响增产措施的效果。

2.5.1 胶囊酸(Encapsulated Acids)

将酸与稠化剂混合,用油或聚合物包裹,作用于裂缝起刻蚀缝面作用。(Gonzalez 等,2000,2001)。

2.5.2 自生酸

在酸压裂广泛应用于碳酸盐岩储层的时候,由于土酸的低岩石溶蚀率,在砂岩地层中的酸压裂还没用于现场。目前用于砂岩有效酸压裂的方法和酸液配方已经研发出来。这种酸压裂的方法包括(Qu 和 Wang,2010):

(1)在足够压开地层的压力下,向地层注入含有磺酸酯、氟化物、支撑剂和水的酸压裂液,磺酸酯水解生成磺酸。

(2)在注入地层酸压裂液后,磺酸与氟化物在地层中发生反应就地生成氢氟酸。

一般情况下,配合局部单层有效支撑以增加支撑裂缝面积,氢氟酸就地反应使砂岩酸压裂成为可能。较宽的裂缝宽度可以增加裂缝导流能力,从而获得比常规加砂压裂更高的产量。

2.5.3 液体滤失

液体滤失限制了酸压裂的效益。因此研发了控制液体滤失的配方(Sanford 等,1992;White 等,1992)。研究表明,增加酸液的黏度可以很好地控制酸压裂液的滤失。在极低渗石灰岩岩心中这种增加是极其明显的。增黏剂的性能也影响着能够成功控制液体滤失,聚合物材料的效果明显好于表面活性剂类增黏剂(Gdanski,1993)。

黏度控制酸中含有凝胶,在泵注完成 1d 后破胶恢复到酸液的初始黏度。这种酸既可以用于基质酸化,也可以用于酸压裂以获得较长的裂缝。酸液的 pH 值控制着酸液的成胶和破胶。适用于 50~135℃的地层温度(Yeager 和 Shuchart,1997)。

2.5.4 酸压裂液的凝胶破胶剂

一种用于钛或锆化合物交联的酸凝胶颗粒破胶剂是由氟化物、磷酸盐、硫酸根离子和多羧酸基化合物等复合材料组成。在颗粒表面涂覆水不溶性树脂涂料,以减少颗粒中破胶剂的释放速率,这样凝胶的黏度降低较为缓慢(Boles 等,1996)。

2.6 特殊问题

2.6.1 缓蚀剂

水溶性 1,2-二硫酚-3-硫酮可以在水相条件下作为压裂液和其他修井液的缓蚀剂(Oude Alink,1993)。这些化合物是异丙苯酚基聚氧化乙烯与硫元素反应制备的。

在水环境条件下,该化合物的性能优于其非草酸盐的衍生物。使用浓度根据体系中水量来确定,一般是 10~500mg/L。

2.6.2 压裂中的铁离子控制

实验室和现场工作都表明,在增产措施的实施过程中,铁离子的存在是个关键并且复杂的问题(Smolarchuk 和 Dill,1986)。在酸液中出现的问题与非酸或弱酸压裂液体系的问题是不同的。一般情况下,酸液会溶解设备和管路中的铁化合物,并且混合后注入地层。

与地层反应一样,酸会溶解很多的铁。如果流体中没有有效的铁离子控制系统,溶解的铁会沉淀。在往井筒返排时,这种沉淀会沉积起来。这种沉积的固体会减少自然和人工的渗透率,并且对处理液的返排和生产有不利影响。

铁可以用某些络合剂进行控制,如葡萄糖酸-δ-内酯、柠檬酸、乙二胺四乙酸、次氨基三乙酸、羟乙基乙二胺三乙酸、羟乙基乙二胺三乙酸及其盐。这些化合物必须与羟胺盐或肼盐等含氮化合物协同使用(Dil 等,1988;Frenier,2001;Walker 等 1987)。图 2.2 给出了部分铁离子络合剂结构式。

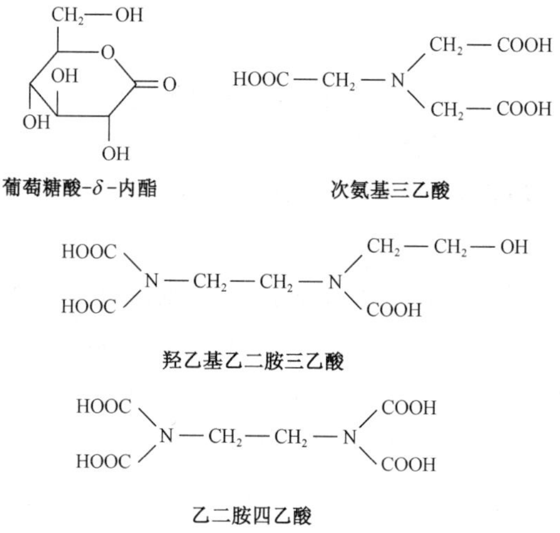

图 2.2 铁离子络合剂

通常,络合剂具有独特的化学特性。这些化合物最重要的特征就是,在水溶液中具有高的酸溶性能。人们用线性岩心驱替试验来研究地层蚓孔。

在 65℃(150°F)的试验条件下,羟乙基乙二胺三乙酸和羟乙基亚氨基二乙酸都会在石灰岩岩心中产生蚓孔。然而其效率和能力是不同的。因为这些化学物质在酸性 pH 值范围内具有高溶解性,可以测试 pH 值小于 3.5 的酸性配方(Frenier 等,2001)。

为了控制水基压裂液中的铁离子,pH 值要低于 7.5,可以使用巯基烷基酸(Brezinski 等,1994)。与前一段中描述的络合剂相反,这是一种铁离子还原剂。

2.6.3 温度稳定剂

人们不希望在压裂施工初期压裂液降解而致使黏度降低。这就需要评价在高温条件下压裂液中聚合物的降解。

一种方法是在压裂前期用大量前置液给地层降温,以防止聚合物过早降解。此外,还可以在泵注前,加入大量未水化的颗粒状瓜尔胶及其衍生物以延长压裂液的温度稳定性(Nimerick 和 Boney,1992)。最后,可以通过调节 pH 值为弱碱性以改善稳定性能。

可以通过硅烷化改善瓜尔胶的耐温性能(Zhang 和 Chen,2012)。最佳合成温度为 85℃,瓜尔胶和氯化三甲基硅烷的物质的量比为 5∶1。硅烷基化瓜尔胶水冻胶的黏度得到很大改进,耐温最高甚至可以达到 80℃。

锆化合物是用于高温的较好的交联剂。表 2.3 给出了高温瓜尔胶压裂液配方。压裂液具有较好的黏度,并且在 80~120℃条件下性能稳定。

表 2.3 高温瓜尔胶压裂液配方(Brannon 等,1989)

组分	作用
瓜尔胶	增黏剂
锆或铪化合物	交联剂
碳酸氢盐	缓冲溶液

2.6.4 化学发泡

可以通过加入发泡剂提高压裂液的返排率(Abou-Sayed 和 Hazlett,1989;Jennings,1995)。加入发泡剂(例如聚团的颗粒和含有发泡剂的颗粒)并压入地层后,发泡剂将会分解,使滤饼变得更加多孔,或提供了将液体从基质中返排出来的驱动力。表 2.4 给出了常用的发泡剂。

表 2.4 发泡剂(Jennings,1995)

化合物	化合物
二亚硝基次戊基四铵	P,P'-双苯磺酰肼
偶氮二甲酰胺	对甲苯磺酰肼
碳酸氢钠	

增加了孔隙也就增加了地层与人工裂缝的连通性能,从而增加了压裂液的返排效率。基质内的气体解吸,增加了人工裂缝与井筒之间的沟通。

2.6.5 防冻配方

表 2.5 给出了一个防冻配方。该体系防冻温度达到了 $-40 \sim -35$℃(Barsukov 等,1993)。

表 2.5 水基压裂液防冻配方(Barsukov 等,1993)

组分	比例(%)
烃类[①]	2~20
表面活性剂	—
矿化水	—
磺酸盐添加剂生产副产品污泥(10%~30%烃类,20%~30%磺酸钙,18%~40%碳酸钙和氢氧化钠)[②]	10~35
乳化剂[③]	0.5~2.0

[①] 气体冷凝物,油或苯。
[②] 降低滤失,提高携砂、抗冻和稳定性能。
[③] 表面活性剂乳化剂。

2.6.6 气井中的地层伤害

低渗透砂岩岩心人工裂缝对地层伤害的研究(Gall 等,1988)表明黏性压裂液限制了气体在较窄裂缝中的流动。聚糖类聚合物,如瓜尔胶、HEC 和黄胞胶会使气体流经破裂岩心的速度明显降低,最大伤害率能达到 95%。

相反,返排后 PAM 凝胶较少或没有降低对气体通道的影响。另一种类型的压裂液,如表面活性剂和破胶剂,对气体流动的伤害也较小。

2.7 压裂液的表征

从历史上看,黏度测量是石油生产中表征流体性能的一个重要方法。很久以前就已经可以在实验室测量液体的流动阻力,而现场对液体性能测量的需求促使了便携式和不太复杂的黏度测量设备的发展(Parks 等,1986)。

这些工具必须满足耐用和操作足够简单,可供技术技能相差很大的人员使用。因此马氏漏斗和两速 Fann 氏同心圆筒黏度计在现场得到了广泛应用。在某些情况下,也有使用 Brookfield 黏度计的。

可以确认严格控制某些变量可以提高水力压裂作业的执行和增产措施的成功。因此提出了严格的质量控制体系(Ely,1996;Ely 等,1990)。这一控制程序包括可监测低温下破胶剂的性能,测量压裂液在较高温度下不同交联剂、温度稳定剂和其他添加剂条件下的交联状态。

2.7.1 流变特征

为了设计一个成功的水力压裂液用交联冻胶,必须精确测量这些流体的流变学特性。以往用旋转黏度计测量硼交联冻胶的流变特征是困难的。现在,可以在实验室装置中,模拟现场泵注条件(例如在线交联)、流体在油管或套管和裂缝中的流动(Shah 等,1988)。可以测量液体的 pH 值、温度以及稠化剂类型和浓度对流体流变性能的影响。

这些参数对裂缝中冻胶的最终黏度有显著影响。摩阻的估算,已经从统计现场不同尺寸管路的摩阻压力,发展到可以从实验室测试得到。与现场比对相结合,这些统计数据可以帮助准确预测硼交联流体的摩阻压力。

2.7.2 锆交联剂

可以测定浓缩的含锆交联剂冻胶中锆离子浓度。首先是添加酸破坏冻胶,将锆转化为非络合的离子形式(Chakrabarti 和 Marczewski,1990)。然后加入偶氮胂(Ⅲ)产生有色络合物,用标准比色法测定。砷化合物是剧毒的。可以用如图 2.3 所示的比色剂测定冻胶中的锆。

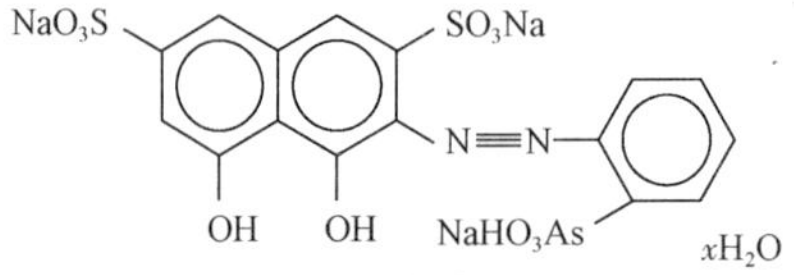

图 2.3 测定冻胶中锆的比色剂

2.7.3 氧化破胶剂

在间歇式或连续取样过程中,可以用比色法测定氧化破胶剂的浓度(Chakrabarti 等,1988)。比色剂含有铁离子和硫氰酸盐,对于氧化剂极为敏感。因此,加入压裂液的破胶剂量是可以控制的。该方法基于亚铁离子氧化为三价铁离子的反应:

$$Fe^{2+} \longrightarrow Fe^{3+} + e^-$$

其与硫氰酸盐形成深红色络合物。

2.7.4 体积排出色谱法

已经成功使用体积排出色谱法(Brannon 和 Tjon,1995;Gall 和 Raible,1986)监测各种氧化和酶破胶剂造成的稠化剂降解。

研究表明,部分破胶的或未破胶的聚合物会显著减少通过多孔介质的流量。瓜尔胶聚合物降解过程中会产生不溶性残渣。这种残留物可以影响介质的孔径(Kyaw 等,2012)。

2.7.5 支撑剂评估

用于支撑剂效果评价的标准化方法为(ISO 13503-5—2006,API RP 19C—2008)。有一般评估方法的实例(Wen 等,2007)。

对于某些支撑剂,实验表明长期裂缝导流能力和闭合压力为多项式的关系(Wen 等,2007)

$$F = A_1 + A_2 p + A_3 p^2 + A_4 p^3 \tag{2.1}$$

其中,F 为长期导流能力,p 为闭合压力,A_i 为常数。类似的,在某些闭合压力下,裂缝导流能力 F 与时间 t 为指数关系(Wen 等,2007),

$$F = \exp(-A_5 t) + A_6 \tag{2.2}$$

文献中的商品名称

商品名	描述	供应商
WS-44	乳化剂(Welton 等,2010)	Halliburton Energy Services,Inc.

参 考 文 献

Abou-Sayed,I. S. ,Hazlett,R. D. ,1989. Removing fracture fluid via chemical blowing agents. US Patent 4 832 123, assigned to Mobil Oil Corp. ,23 May 1989.

Allan,T. L. ,Amin,J. ,Olson,A. K. ,Pierce,R. G. ,2008. Fracturing fluid containing amphoteric glycinate surfactant. US Patent 7 399 732,assigned to Calfrac Well Services Ltd. ,Calgary,Alberta,CA,Chemergy Ltd. ,Calgary,Alberta,CA,15 July 2008. <http://www. freepatentsonline. com/7399732. html>.

Anonymous,1999. Fracturing products and additives. World Oil 220(8),135,137,139-145.

API Standard RP 19C,2008. Recommended practice for measuring the long-term conductivity of proppants. API Standard API RP 19C,American Petroleum Institute,Washington,DC.

Barsukov,K. A. ,Ismikhanov,V. Y. ,Akhmetov,A. A. ,Pop,G. S. ,Lanchakov,G. A. ,Sidorenko,V. M. ,1993. Composition for hydro-bursting of oil and gas strata—consists of hydrocarbon phase,sludge from production of sulphonate additives to lubricating oils,surfactant-emulsifier and mineralised water. SU Patent 1 794 082,assigned to Urengoi Prod. Assoc. ,7 February 1993.

Boles,J. L. ,Metcalf,A. S. ,Dawson,J. C. ,1996. Coated breaker for crosslinked acid. US Patent 5 497 830,assigned to BJ Services Co. ,12 March 1996.

Bonekamp,J. E. ,Rose,G. D. ,Schmidt,D. L. ,Teot,A. S. ,Watkins,E. K. ,1993. Viscoelastic surfactant based foam fluids. US Patent 5 258 137,assigned to Dow Chemical Co. ,2 November 1993.

Brannon,H. D. ,Tjon,J. P. R. M. ,1995. Characterization of breaker efficiency based upon size distribution of polymeric fragments. In:Proceedings Volume Annual SPE Technical Conference,Dallas,22-25 October 1995,pp. 415-429.

Brannon, H. D., Hodge, R. M., England, K. W., 1989. High temperature guar - based fracturing fluid. US Patent 4 801 389, assigned to Dowell Schlumberger Inc., 31 January 1989.

Brezinski, M., Gardner, T. R., Harms, W. M., Lane Jr., J. L., King, K. L., 1994. Controlling iron in aqueous well fracturing fluids. EP Patent 599 474, assigned to Halliburton Co., 1 June 1994.

Chakrabarti, S., Marczewski, C. Z., 1990. Determining the concentration of a cross - linking agent containing zirconium. GB Patent 2 228 996, assigned to British Petroleum Co. Ltd., 12 September 1990.

Chakrabarti, S., Martins, J. P., Mealor, D., 1988. Method for controlling the viscosity of a fluid. GB Patent 2 199 408, assigned to British Petroleum Co. Ltd., 6 July 1988.

Chatterji, J., King, B. J., King, K. L., 2007. Recyclable foamed fracturing fluids and methods of using the same. US Patent 7 205 263, assigned to Halliburton energy Services, Inc., Duncan, OK, 17 April 2007. < http://www.freepatentsonline.com/7205263.html >.

Cottrell, T. L., Spronz, W. D., Weeks III, W. C., 1988. Hugoton infill program uses optimum stimulation technique. Oil Gas J. 86(28), 88 - 90

Deng, Q., Xu, J., Gu, X., Tang, Y., 2012. Properties evaluation of polymer - free fluid for fracturing application. Adv. Mater. Res. 482 - 484 (Pt. 2, Advanced Composite Materials), 1180 - 1183. http://dx.doi.org/10.4028/www.scientific.net/AMR.482 - 484.1180.

Dill, W. R., Ford, W. G. F., Walker, M. L., Gdanski, R. D., 1988. Treatment of iron - containing subterranean formations. EP Patent 258 968, 9 March 1988.

Ely, J. W., 1996. How intense quality control improves hydraulic fracturing. World Oil 217(11), 59 - 60, 62 - 65, 68.

Ely, J. W., Wolters, B. C., Holditch, S. A., 1990. Improved job execution and stimulation success using intense quality control. In: Proceedings Volume 37th Annual Southwestern Petroleum Short Course Association et al Meeting, Lubbock, Texas, 18 - 19 April 1890, pp. 101 - 114.

Frenier, W. W., 2001. Well treatment fluids comprising chelating agents. WO Patent 0 183 639, assigned to Sofitech NV, Schlumberger Serv. Petrol, Schlumberger Canada Ltd., Schlumberger Technol. BV, and Schlumberger Holdings Ltd., 8 November 2001.

Frenier, W. W., Fredd, C. N., Chang, F., 2001. Hydroxyaminocarboxylic acids produce superior formulations for matrix stimulation of carbonates. In: Proceedings Volume SPE EuropeFormation Damage Conference, the Hague, Netherlands, 21 - 22 May 2001).

Gall, B. L., Raible, C. J., 1986. The use of size exclusion chromatography to study the degradation of water - soluble polymers used in hydraulic fracturing fluids. In: Proceedings 192nd ACS National Meetings, vol. 55. American Chemical Society Polymeric Materials: Science and Engineering Division Technology Program, Anaheim, Calif, 7 - 12 September 1986, pp. 572 - 576.

Gall, B. L., Maloney, D. R., Raible, C. J., Sattler, A. R., 1988. Permeability damage to natural fractures caused by fracturing fluid polymers. In: Proceedings Volume SPE Rocky Mountain Regional Meeting, Casper, Wyo, 11 - 13 May 1988, pp. 551 - 560.

Gdanski, R. D., 1993. Fluid properties and particle size requirements for effective acid fluid - loss control. In: Proceedings Volume SPE Rocky Mountain Regional Meeting: Low Permeability Reservoirs Symposium, Denver, 26 - 28 April 1993, pp. 81 - 94.

Gonzalez, M. E., Looney, M. D., 2000. The use of encapsulated acid in acid fracturing treatments. WO Patent 0 075 486, assigned to Texaco Development Corp., 14 December 2000.

Gonzalez, M. E., Looney, M. D., 2001. Use of encapsulated acid in acid fracturing treatments. US Patent 6 207 620, assigned to Texaco Inc., 27 March 2001.

Harris, P. C., 1988. Fracturing - fluid additives. J. Pet. Technol. 40(10), 1277 - 1279.

Holditch, S. A., Xiong, H., Rahim, Z., Rueda, J., 1993. Using an expert system to select the optimal fracturing fluid and treatment volume. In: Proceedings Volume SPE Gas Technology Symposium, Calgary, Canada, 28 – 30 June 1993, pp. 515 – 527.

Hrachovy, M. J., 1994a. Hydraulic fracturing technique employing in situ precipitation. WO Patent 9 406 998, assigned to Union Oil Co. California, 31 March 1994.

Hrachovy, M. J., 1994b. Hydraulic fracturing technique employing in situ precipitation. US Patent 5 322 121, assigned to Union Oil Co. California, 21 June 1994.

ISO – 13503 – 5, 2006. Petroleum and natural gas industries—completion fluids andmaterials—Part 5: Procedures formeasuring the long – term conductivity of proppants. ISO Standard ISO – 13503 – 5, International Organization for Standardization, Geneva, Switzerland.

Jennings, Jr., A. R., 1995. Method of enhancing stimulation load fluid recovery. US Patent 5 411 093, assigned to Mobil Oil Corp., 2 May 1995.

Kyaw, A., Nor Azahar, B. S., Tunio, S. Q., 2012. Fracturing fluid (guar polymer gel) degradation study by using oxidative and enzyme breaker. Res. J. Appl. Sci. Eng. Technol. 4(12), 1667 – 1671. < http://maxwellsci.com/print/rjaset/v4 – 1667 – 1671. pdf >.

Lukocs, B., Mesher, S., Wilson, T. P. J., Garza, T., Mueller, W., Zamora, F., Gatlin, L. W., 2007. Non – volatile phosphorus hydrocarbon gelling agent. US Patent Application 20070173413, assigned to Clearwater International, LLC, 26 July 2007. < http://www.freepatentsonline.com/20070173413.html >.

Middaugh, R. L., Harris, P. C., Heath, S. J., Taylor, R. S., Hoch, O. F., Phillippi, M. L., Slabaugh, B. F., Terracina, J. M., 2007. Coarse – foamed fracturing fluids and associated methods. US Patent 7 261 158, assigned to Halliburton Energy Services, Inc., Duncan, OK, 28 August 2007. < http://www.freepatentsonline.com/7261158.html >.

Mukherjee, H., Cudney, G., 1992. Extension of acid fracture penetration by drastic fluid – loss control. SPE Unsolicited Pap.

Nimerick, K. H., Boney, C. L., 1992. Method of fracturing high temperature wells and fracturing fluid therefore. US Patent 5 103 913, assigned to Dowell Schlumberger Inc., 14 April 1992.

Oude Alink, B. A., 1993. Water soluble 1,2 – dithio – 3 – thiones. US Patent 5 252 289, assigned to Petrolite Corp., 12 October 1993.

Parks, C. F., Clark, P. E., Barkat, O., Halvaci, J., 1986. Characterizing polymer solutions by viscosity and functional testing. In: Proceedings of the 192nd ACS National: Meeting, vol. 55. American Chemical Society Polymeric Materials: Science and Engineering Division Technology Program, Anaheim, Calif, 7 – 12 September 1986, pp. 880 – 888.

Penny, G. S., Conway, M. W., 1989. Fluid Leakoff. Recent Advances In Hydraulic Fracturing (SPE Henry L. Doherty Monogr Ser), vol. 12. SPE, Richardson, Texas, pp. 147 – 176.

Qu, Q., Wang, X., 2010. Method of acid fracturing a sandstone formation. US Patent 7 704 927, assigned to BJ Services Co., Houston, TX, 27 April 2010. < http://www.freepatent sonline.com/7704927.html >.

Sanford, B. D., Dacar, C. R., Sears, S. M., 1992. Acid fracturing with new fluid – loss control mechanisms increases production, little knife field, North Dakota. In: Proceedings Volume SPE Rocky Mountain Regional Meeting, Casper, Wyo, 18 – 21 May 92, pp. 317 – 324.

Shah, S. N., Harris, P. C., Tan, H. C., 1988. Rheological characterization of borate crosslinked fracturing fluids employing a simulated field procedure. In: Proceedings Volume SPE Production Technology Symposium, Hobbs, New Mexico, 7 – 8 November 1988.

Smolarchuk, P., Dill, W., 1986. Iron control in fracturing and acidizing operations. In: Proceedings Volume 37thAnnual CIM Petroleum Society Technical Meeting, Calgary, Canada, vol. 1, 8 – 11 June 1986, pp. 391 – 397.

Stacy, A. L., Weber, R. B., 1995. Method for reducing deleterious environmental impact of subterranean fracturng processes. USPatent 5 424 285, assigned to WesternCo. NorthAmerica, 13 June 1995.

Tamayo, H. C., Lee, K. J., Taylor, R. S., 2008. Enhanced aqueous fracturing fluid recovery from tight gas formations:

Foamed CO2 pre – pad fracturing fluid and more effective surfactant systems. J. Can. Pet. Technol. 47(10), 33 – 38. http://dx. doi. org/10. 2118/08 – 10 – 33.

Walker, M. L. , Ford, W. G. F. , Dill, W. R. , Gdanski, R. D. , 1987. Composition and method of stimulating subterranean formations. US Patent 4 683 954, 4 August 1987.

Wang, Z. , Wang, S. , Sun, X. , 2012. The influence of surfactant concentration on rheology and proppant – carrying capacity of VES fluids. Adv. Mater. Res. 361 – 363 (Pt. 1, Natural Resources and Sustainable Development), 574 – 578. http://dx. doi. org/10. 4028/www. scientific. net/AMR. 361 – 363. 574.

Welton, T. D. , Todd, B. L. , McMechan, D. , 2010. Methods for effecting controlled break in pH dependent foamed fracturing fluid. US Patent 7 662 756, assigned to Halliburton Energy Services, Inc. , Duncan, OK, 16 February 2010. < http://www. freepatentsonline. com/7662756. html >.

Wen, Q. , Zhang, S. , Wang, L. , Liu, Y. , Li, X. , 2007. The effect of proppant embedment upon the long – term conductivity of fractures. J. Pet. Sci. Eng. 55 (3 – 4), 221 – 227. < http://www. science. direct. com/science/article/B6VDW – 4M6459G – 2/2/977b6f4fd9756baba71ba40771815d40 >.

White, D. J. , Holms, B. A. , Hoover, R. S. , 1992. Using a unique acid – fracturing fluid to controlfluid loss improves stimulation results in carbonate formations. In: Proceedings Volume SPE Permian Basin Oil and Gas Recovery Conference, Midland, Texas, 18 – 20 March 1992, pp. 601 – 610.

Xiong, H. , Davidson, B. , Saunders, B. , Holditch, S. A. , 1996. A comprehensive approach to select fracturing fluids and additives for fracture treatments. In: Proceedings Volume Annual SPE Technical Conference, Denver, 6 – 9 October 1996, pp. 293 – 301.

Yeager, V. , Shuchart, C. , 1997. In situ gels improve formation acidizing. Oil Gas J. 95(3), 70 – 72.

Zhang, J. , Chen, G. , 2012. Improvement of the temperature resistance of guar gum by silanization. Adv. Mater. Res. 415 – 417 (Pt. 1, Advanced Materials), 652 – 655. http://dx. doi. org/10. 4028/www. scientific. net/AMR. 415 – 417. 652

3 稠 化 剂

多种化合物均可作为稠化剂使用,如表 3.1 所示。本章对这些化合物进行详细介绍。

表 3.1 稠化剂

化合物	参考文献
亲水性和疏水性单体的水溶性共聚物,硅烷或硅氧烷的丙烯酰胺(AM)-丙烯酸酯	Meyer 等(1999)
羧甲基纤维素,聚乙二醇	Lundan 等(1993),Lundan 和 Lahteenmaki(1996)
纤维素醚和黏土的化合物	Rangus 等(1993)
酰胺改性并含有羧基的聚合糖类	Batelaan 和 van derHorts(1994)
铝酸钠和氧化镁	Patel(1994)
具有热稳定性的 30% 的羟乙基纤维素(HEC)硫代硫酸钠或硫代硫酸铵和 20% 的 HEC	Lukach 和 Zapico(1994)
丙烯酸(AA)共聚物和与疏水基氧化烯	Egraz 等(1994)
丙烯酰胺-丙烯酸酯的共聚物和乙烯基磺酸盐-乙烯基酰胺的共聚物	Waehner(1990)
阳离子聚半乳聚甘露糖和阴离子黄胞胶	Yeh(1995b)
乙烯基聚氨酯和 AA 或烷基丙烯酸酯的共聚物	Wilkerson 等(1995)
2-硝化烷基醚化改性淀粉	Gotlieb 等(1996)
葡萄糖醛醛酸聚合物	Courtois-Sambourg 等(1993)
铬铁木质素磺酸盐和羧甲基纤维素	Kotelnikov 等(1992)
纤维素纳米原纤维[①]	Langlois(1998) 和 Langlois 等(1999)
四元烷基酰胺季铵盐	Subramanian 等(2001a)
聚氨基葡糖[②]	House 和 Cowan(2001)

① 热稳定性达到 180℃。
② 可溶于酸液。

(1)聚合物。

增稠性聚合物包括聚氨酯、聚酯和聚丙烯酰胺,除此之外还有天然聚合物以及改性天然聚合物(Doolan 和 Cody,1995)。

(2)pH 值关联型稠化剂。

离子型聚合物的黏度与 pH 值相关,特别是 pH 值相关型稠化剂,此类稠化剂可以通过丙烯酸、甲基丙烯酸乙基丙烯酸酯或其他乙烯单体和三苯乙烯聚(氧乙烯)$_x$ 甲基丙烯酸酯共聚得到。这种共聚物在 pH 值低于 5.0 的酸性环境中是稳定的水胶态分散体,但是当 pH 值达到

5.5~10.5或更高时,在水溶液中就会成为一种有效的稠化剂(Robinson 1996a,1999b)。

(3)混合金属氢氧化物。

典型的膨润土钻井液中添加混合金属氢氧化物,可转变为一种剪切变稀的流体(Lange和Plank,1999),静止时这些流体黏度较高,施加剪切应力时就会变得像水一样稀。

理论上,混合金属氢氧化物和膨润土混合后形成了三维、脆弱的网状结构,因此流体才具有剪切变稀的流变行为。

带正电的混合金属氢氧化物颗粒能够附着于表面带负电荷的膨润土薄片上。氢氧化镁铝盐就是混合金属氢氧化物。

在钻井过程中,混合金属氢氧化物存在以下优势(Felixberger,1996):
① 具有较高的清除岩屑能力;
② 在关井时悬浮固体颗粒;
③ 降低泵的摩擦阻力;
④ 稳定井眼;
⑤ 提高钻井速率;
⑥ 保护产出层段。

混合金属氢氧化物钻井液已经成功用于连续油管水平井钻井,也用于在河流、公路、港湾下挖掘隧道时挖掘更大孔洞,能够起到扩冲孔眼、巩固管道的作用。

用铵处理过的相应氯化物可用于制备混合金属氢氧化物(Burba和Strother,1991)。多种钻井液实验表明:混合金属氢氧化物体系配合丙二醇(Deem等,1991)使用,表皮伤害最低。

由天然矿物,特别是铝碳酸镁制成的混合金属氢氧化物,这种氢氧化物可采用热能激活,因为除镁、铝外可能还含有少量或微量具有激活作用的金属杂质(Keilhofer和Plank,2000)。

文献中还介绍了具有石榴石型三维空间晶格结构的二价和三价金属混合氢氧化物[$Ca_3Al_2(OH)_{12}$](Burba等,1992;Mueller等,1997)。

3.1 水基压裂液用稠化剂

稠化剂也称增黏剂,指的是能够使压裂液变成凝胶,从而增加其黏度的化合物(Welton等,2010)。

适合做稠化剂的聚合物有:瓜尔胶、黄胞胶、维纶胶、槐豆胶、印度树胶、刺梧桐胶、罗望子胶、黄芪胶。瓜尔胶可以进行功能化改性,如羟乙基瓜尔胶、羟丙基瓜尔胶以及羧甲基瓜尔胶。水溶性纤维素醚类包括甲基纤维素、羧甲基纤维素(CMC)、羟乙基纤维素(HEC)以及羟乙基羧甲基纤维素(Welton等,2010)。

合成聚合物也可作为稠化剂使用,如丙烯酰胺、甲基丙烯酰胺、丙烯酸和甲基丙烯酸的共聚物、2-丙烯酰氨基-2-甲基-1-丙烷磺酸衍生物(AMPS)和N-乙烯基吡啶的共聚物(Welton等,2010),除此之外还有天然多糖及其衍生物(Lemanczyk,1992)。这些聚合物用量很少即可增加液体黏度。适合用于压裂液的聚合物见表3.2。

表 3.2 压裂液用稠化剂

稠化剂	参考文献
羟丙基瓜尔胶①	
半乳甘露聚糖②	Mondshine(1987)
羟乙基纤维素改性的乙烯膦酸	Holtmyer and Hunt(1992)
羧甲基纤维素	
N-乙烯基内酰胺单体合成的聚合物乙烯基磺酸盐③	Bharat(1990)
网状细菌纤维素④	Westland et al.(1993)
细菌黄胞胶⑤	Hodge(1997)

① 与淀粉相比,增稠能力高 8 倍。
② 用硼交联剂进行交联,能够增加其耐温性。
③ 具有高温稳定性。
④ 优良的液体性能。
⑤ 黏度更高。

瓜尔胶结构如图 3.1 所示,对一些羟基进行丙烷基醚化,形成羟丙基瓜尔胶。油基压裂液用稠化剂与水基压裂液用稠化剂有所不同,稠化剂可能由以下化合物构成:磷酸酯、交联剂、多价金属离子、某种催化剂或某种饱和四元胺(Lawrence 和 Warrender 等,2010)。

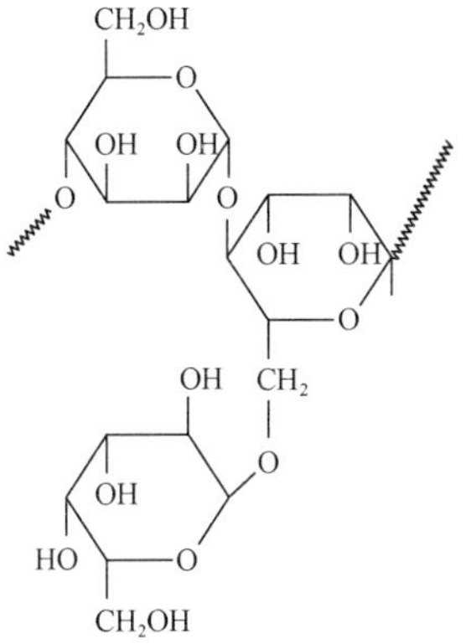

图 3.1 瓜尔胶结构单元

锆交联剂构成:

商用锆交联剂,有的在高 pH 值条件下交联过快,如三乙醇胺锆,冻胶不耐剪切(Putzig,2012);也有的交联速度太慢(Putzig,2012),如三乙醇胺锆化合物在剪切作用下,会有黏度损失(Rummo 和 Startup,1986;Kucera,1987;Baranet 等,1987)。

水基压裂液由以下组分构成:pH 缓冲剂、可交联的有机聚合物、锆交联剂(Putzig,2012)。锆交联剂配位体有链烷醇胺和乙二醇。

常用于合成锆交联剂的四烷基锆酸盐是正丙醇锆酸酯,如 TYZOR NPZ®,是一种丙醇溶液,其中 ZrO_2 的质量分数约 28%。

压裂液组分中包括可交联的有机聚合物,如瓜尔胶衍生物。然而,在压裂液体系 pH 值低于 6.0 或高于 9.0 时,或为了减少固体残渣,降低储层伤害,采用纤维素压裂液更具优势(Putzig,2012)。

3.1.1 瓜尔胶

瓜尔豆是瓜尔豆属的一种多分支糖,起源于印度,目前美国南部也有种植。其分子质量大约220kDa❶,由甘露糖主链和半乳糖支链构成,二者比例是2:1。具有这种结构的多糖被称为异甘露聚糖,特别是半乳甘露聚糖,因此瓜尔胶衍生物有时也被称为半乳甘露聚糖。

由于瓜尔胶类稠化剂具有良好的流变性、经济性和水溶性,特别是羟丙基瓜尔胶,因此被广泛用作压裂液稠化剂。非乙酰化黄胞胶是一种改性黄胞胶,能够与瓜尔胶产生协同作用,在低聚合物浓度下即可获得更好的黏度和支撑剂悬浮能力。

静态滤失实验表明:硼交联与锆交联羟丙基瓜尔胶压裂液的滤失系数相同(Zeilinger等,1991)。应力敏感性实验表明:锆交联压裂液滤饼具有黏弹性,但硼交联压裂液滤饼仅具有弹性。大量岩心流动实验表明:在较宽的岩心渗透率范围内,非交联液体没有滤饼行为,而是一种黏性流动,并且这种流动依赖于多孔介质的特性。

在水基压裂液中加入二醇类物质,如乙二醇(EG),能与瓜尔胶产生胶凝作用,增加液体黏度,稳定液体中的盐类,在27~177℃(80~350°F)条件下非常稳定,并且能够使压裂后的储层伤害降到最低。与瓜尔胶聚合物相比,这种液体通过添加乙二醇可以获得同样的黏度,且伤害更小(Kelly等,2007)。

交联剂可以是硼酸、钛酸或锆酸盐。压裂液中加入硫代硫酸钠有利于提高冻胶稳定性。在93℃(200°F)条件下,在5%的氯化钾、2.4kg/m³的瓜尔胶溶液中加入不同浓度的EG,实验结果见图3.2。

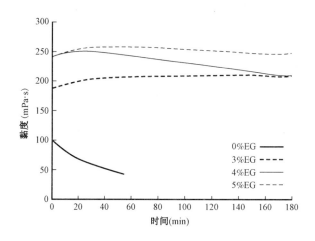

图3.2 次乙基乙二醇(EG)浓度不同时,盐水瓜尔胶溶液的黏度对比(Kelly等,2007)

在含有EG的压裂液中采用乙二胺四乙酸钠作为破胶剂,破胶时间可根据需要而定(Kelly等,2007)。

用AMPS和1-烯丙氧基-2-羟丙基磺酸制得的磺酸盐基团对瓜尔胶羟基进行部分酯化,得到的阴离子半乳甘露聚糖瓜尔胶是一种非常好的稠化剂(Yeh,1995),无论是单独使用还是与其他阳离子聚合物配合使用都会获得较高的黏度。

---

❶ 千道尔顿,1kDa = 1000摩尔质量。

多羟基化合物可通过多种反应进行改性，如以葡萄糖作为改性化合物就是醚化改性，如图3.3所示；采用乙烯基化合物进行瓜尔胶改性，如图3.4所示。

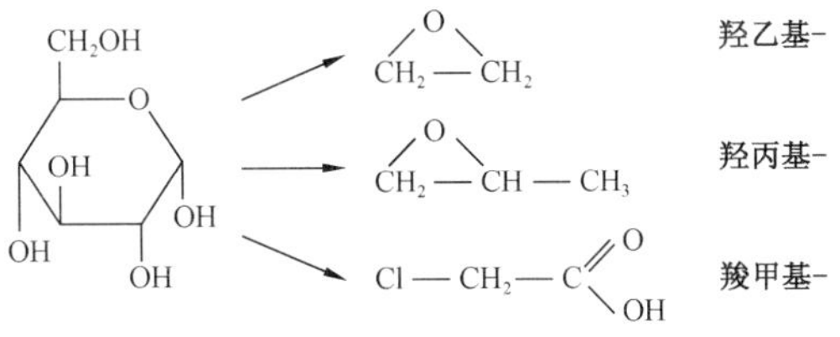

图3.3 多羟基化合物醚化改性

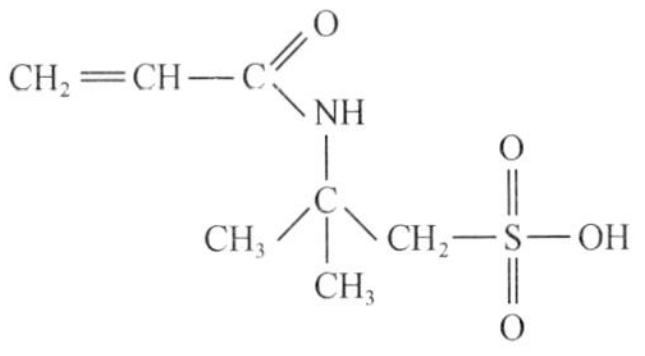

2-丙烯酰胺基-2-甲基-1-丙磺酸

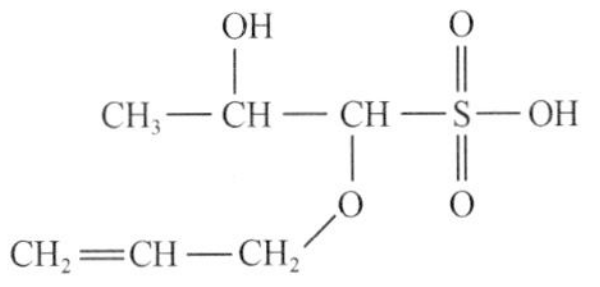

1-烯丙氧基-2-羟丙基磺酸

图3.4 瓜尔胶乙烯基化改性

添加溶解速率较低的少量可溶性硼酸盐，能够提高半乳甘露聚糖压裂液的温度稳定性。随着温度升高，硼酸盐释放出硼离子，从而提高半乳甘露聚糖的冻胶稳定性。

纳米粒子对压裂液流变性能的影响已有研究（Jafry 等，2011）。在气化生成的碳纤维上涂硅，然后再用十八烷基三氯甲硅烷进行功能化处理，把这种改性纤维加入到压裂液冻胶中，分别测试 pH 值为 8.6、9.3 和 10.3 条件下的流变性，结果表明仅在低于常规压裂液 pH 值时改性纤维和冻胶发生反应。

在瓜尔胶压裂液中加入改性纳米粒子对储能模量影响较小，在施加剪切应力时也不会产生永久破坏，这种性能与普通瓜尔胶类似。另外，改性纳米颗粒的存在也不会改变交联瓜尔胶的结构和交联位置（Jafry 等，2011）。

聚烷氧基烯烃基酰胺接枝改性的瓜尔胶是很好的瓜尔胶衍生物。瓜尔胶或羟丙基瓜尔胶的改性需要三个步骤（Bahamdan 和 Daly，2007）：

（1）用氯乙酸钠进行羧甲基化；

（2）用硫酸二甲酯进行酯化；

（3）用聚烷氧基烯烃基酰胺进行酰胺化。

首先进行羟丙基瓜尔胶的羧甲基化，取代度是 0.2 ~ 0.3，再用聚烷氧基烯烃基酰胺进行改性，用红外光谱分析改性后的瓜尔胶，分子质量达到 300 ~ 3000Da。

在聚烷氧基烯烃基酰胺中,可以从9∶1到8∶58调节氧化丙烯与聚氧乙烯的比率,获得不同疏水性的目标产物。接枝瓜尔胶比非接枝瓜尔胶黏度低一到两个数量级(Bahamdan 和 Daly,2007)。

3.1.2 羟乙基纤维素

利用羟乙基纤维素 HEC 与乙烯基膦酸反应对其进行改性,改性过程中会产生过氧化氢和亚铁盐,最终形成 HEC 接枝共聚物。

直链淀粉是一种葡萄糖线性聚合物,可溶于水,与纤维素的区别在于葡萄糖单元连接方式不同,直链淀粉是 α 键链接,而纤维素是 β 键链接。正是这种差异导致直链淀粉可溶于水,而纤维素不溶于水,对纤维素进行化学改性可使其成为水溶性聚合物。直链淀粉和纤维素分子结构如图 3.5 所示。

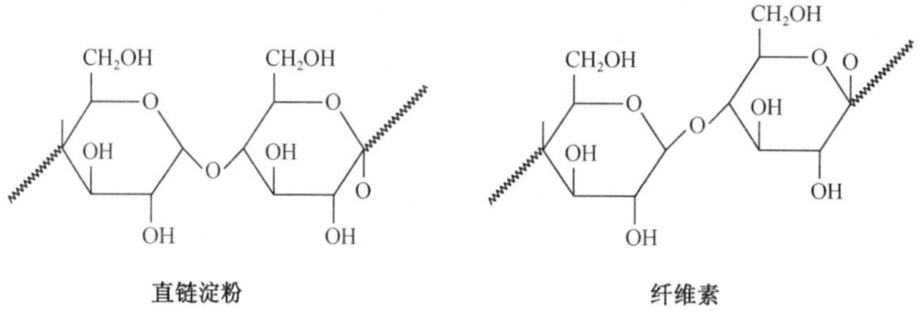

图 3.5 直链淀粉和纤维素

改性羟乙基纤维素 HEC 已经被用作水力压裂液稠化剂(Holtmyer 和 Hunt,1992),并且可采用多价金属阳离子进行交联,增加压裂液黏度。

3.1.3 生物技术产品

3.1.3.1 吉兰胶和维纶胶

吉兰胶的通用名称是假单胞杆菌产生的胞外多糖,是一种线性阴离子多糖,分子质量为 500kDa,包括 1,3 - β - D - 葡萄糖,1,4 - β - D - 葡萄糖醛酸,1,4 - β - D - 葡萄糖和 1,4 - α - L - 甲基戊糖。

维纶胶是通过有氧发酵制得的,其主链与吉兰胶相同,但它有一个侧链 L - 甘露糖或 L - 甲基戊糖。维纶胶可用于控制液体滤失,在碱性环境下与钙离子能够完全配伍。

3.1.3.2 网状细菌纤维素

细菌纤维素由细菌生成,结构错综复杂,与传统的纤维素不同,细菌纤维素具有独特的属性和功能,可以提高水基液体的流变性和粒子悬浮能力(Westland 等,1993)。

3.1.3.3 黄胞胶

黄胞胶由野油菜黄单胞菌诱导制得,自1964年以来就已经投入商业应用了。黄胞胶是一种水溶性多糖聚合物,其重复单元的分子结构见表 3.3 和图 3.6(Doherty 等,1992)。

表 3.3 不同的黄胞胶

数目	重复单元	比例
五聚物	D-葡萄糖∶D-甘露糖∶D-葡萄糖醛酸	2∶2∶1
四聚物	D-葡萄糖∶D-甘露糖∶D-葡萄糖醛酸	2∶1∶1

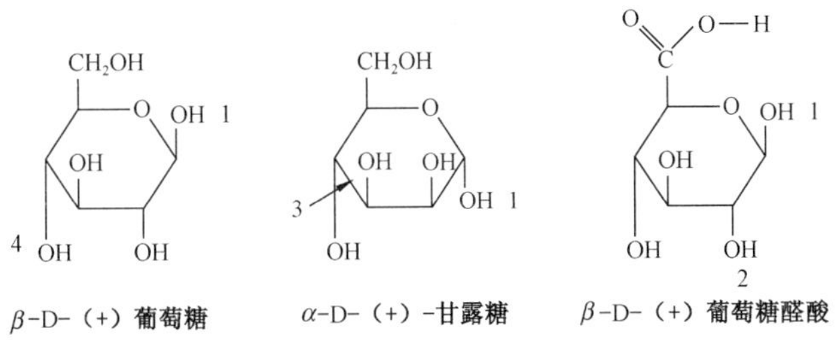

图 3.6 糖类及其衍生物

D-葡萄糖根与 $\beta-(1,4)$ 结构相连,内部 D-甘露糖根与 $\alpha-(1,3)$ 结构相连,一般与葡萄糖根交替出现。D-葡萄糖醛酸与 $\beta-(1,2)$ 结构相连,然后连接到内部甘露糖根上面,外甘露糖与葡萄糖醛酸相连,然后连接到 $\beta-(1,4)$ 结构上面。

在油田的应用中,常用的黄胞胶为含有 8%~15% 聚合物的发酵液,不像其他多糖,其黏度对温度依赖性较小。

3.1.4 黏弹性压裂液配方

与传统聚合物压裂液相比,黏弹性表面活性剂(VES)压裂液体系具有以下优势(Li 等,2010):

(1)保持含油储层的高渗透率;
(2)降低储层伤害;
(3)压裂后增黏剂具有高回收率;
(4)不需要添加酶或氧化破胶剂;
(5)溶解速度快。

VES 压裂液体系缺点在于成本较高,耐盐、耐温性能差,不适合用于深井压裂,目前在这些方面已有所突破。

黏弹性表面活性剂压裂液组成如下:一种两性离子表面活性剂、二十二烯氨基丙烷基甜菜碱和阴离子聚合物或 N-二十二烯-N,N-双(2-羟乙基)-N-甲基氯化铵、聚萘磺酸盐和阳离子表面活性剂,阳离子表面活性剂有甲基聚氧乙烯、十八烷基氯化铵和聚氧乙烯椰油烷基胺(Couillet 和 Hughes,2008;Li 等,2010)。由这些表面活性剂组成的液体黏度较好。

典型的黏弹性表面活性剂是 N-二十二烯-N,N-双(2-羟乙基)-N-甲基氯化铵和油酸钾,与水杨酸钠和氯化钾混合后可形成凝胶(Jones 和 Tustin,2010)。

阳离子型表面活性剂既可溶于有机溶剂,也可溶于无机溶剂。向表面活性剂活性单元上连接多个长链烷基可提高其在有机溶剂中的溶解度(Jones 和 Tustin,2010),例如:十六烷基磷酸三丁酯和三辛烷基甲基胺。

用于配制黏弹性溶液的阳离子表面活性剂链上有一个相当长的线性烃基,显而易见,表面活性剂在有机溶剂中的溶解度和形成黏弹性溶液需求方面存在矛盾。因此需要设计一种适用于水基井筒工作液的增稠剂,既可溶于有机溶剂也可溶于水中的表面活性剂化合物。

牛油酰胺丙胺氧化物(McElfresh 和 Williams,2007)就是一种合适的非离子表面活性剂稠化剂。非离子表面活性剂液体比阳离子表面活性剂液体对储层伤害更小,也比阴离子稠化剂更加有效。

含支链的油酸盐合成过程如下:在嘧啶类催化剂催化作用下,利用甲基油酸和碘化甲烷即可制备 2-甲基油酸甲酯或 2-甲基油酸(Jones 和 Tustin,2010)。

矿物油能够打破 VES 内部结构,可用作黏弹性表面活性剂压裂液破胶剂(Crews 等 2010),在一定温度条件下,破胶速度快慢取决于盐的类型和加量。如果使用低分子矿物油作为破胶剂,则必须在 VES 表面活性剂加入水中之后添加。此类破胶剂可根据需求调节破胶时间。除了矿物油之外还可以采用脂肪酸化合物或细菌作为黏弹性表面活性剂压裂液的破胶剂(Crews,2006;Crews 等,2010)。

3.1.5 其他聚合物

由于酯基疏水,酸基亲水,因此 2″-乙基己基丙烯酸酯和丙烯酸的共聚物既不溶于水也不溶于油。2″-乙基己基丙烯酸酯如图 3.7 所示。在制备压裂液时加入粒径小于 $10\mu m$ 的悬浮物能够提高压裂液性能(Harms 和 Norman,1988)。

在高温储层中,水溶性聚合物 N-乙烯内酰胺或含有乙烯基的磺酸盐单体能够降低滤失,提高压裂液性能(Bharat,1990)。分散剂有褐煤、单宁和沥青。乙烯基单体如图 3.8 所示。

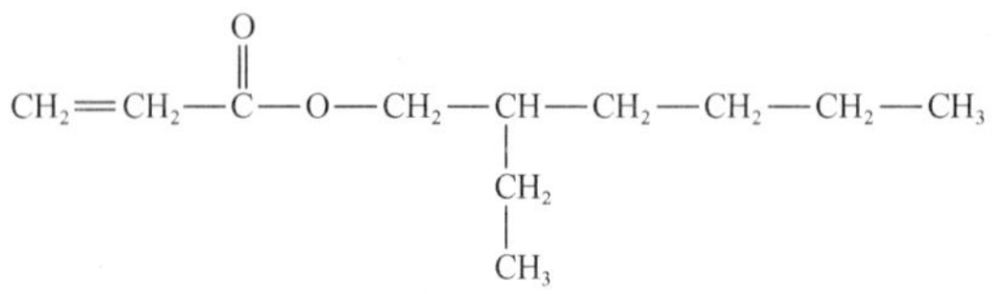

图 3.7 2″-乙基己基丙烯酸酯

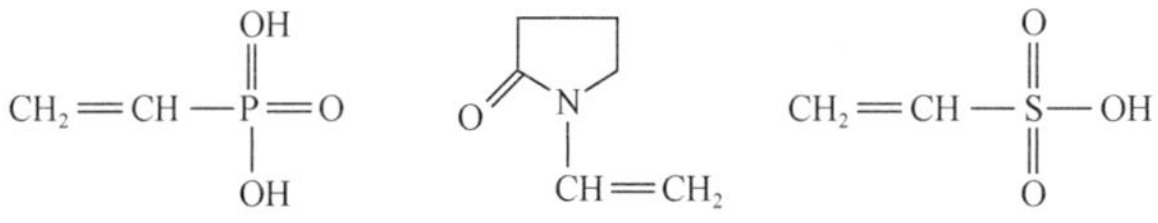

图 3.8 合成稠化剂单体

3.1.5.1 丙交酯聚合物

压裂液稠化剂也可使用可降解的热塑性丙交酯聚合物,其降解的主要机理是水解(Cooke,2009)。

3.1.5.2 可生物降解的配方

可生物降解的钻井液配方中包括一种多糖,其浓度不大,完全可以靠细菌降解。这种聚合物可以是高黏度羧甲基纤维素(CMC),对多糖降解产生的细菌酶比较敏感(Pelissier Biasini,1991)。

通过测定水中溶解氧含量的方法可以研究 7 种钻井液添加剂的生物降解性,这是一种简单生化需氧量测试方法。淀粉生物降解能力高,但低于烯丙基单体合成的聚合物和含芳基的添加剂的降解性(Guo 等,1996)。

3.2 浓缩液

传统的配液技术是在配液罐中提前配制,然后进行循环,直到达到人们期望的流体性能。这种方法不仅耗费时间,而且如果施工提前结束就会增加石油公司的液体处理工作。

如果能够根据需要现场配制液体,则可避免污染环境及后期处理工作,因此研发了连续混配技术,能够配制水基、甲醇和油基液体(Gregory 等,1991)。该技术避免了提前批量配制,减小了添加剂用量和基液处理工作。工作人员只需负责所用液体性能,不再考虑与环境相关的问题,并且可以采用计算机在线程序对化学添加剂进行检测,通过改变聚合物加量获得所需液体流变性能,确保整个过程中的质量控制。

将羟丙基瓜尔胶溶解到柴油中,制成瓜尔胶浓缩液可实现在线连续混配(Harms 等,1988),简化整个施工过程,减轻公司负担。

表 3.4 列出了压裂液用浓缩稠化剂的组分(Brannon,1988)。这种聚合物浓缩液容易分散,在适当的 pH 值条件下,增黏较快,非常适合大规模在线施工。合适的表面活性剂如图 3.9 所示。

表 3.4 浓缩稠化剂组分(Brannon,1988)

组分	示例
疏水基溶剂	柴油
悬浮剂	有机黏土
表面活性剂	乙氧基壬基苯酚
水溶性聚合物	羟丙基瓜尔胶

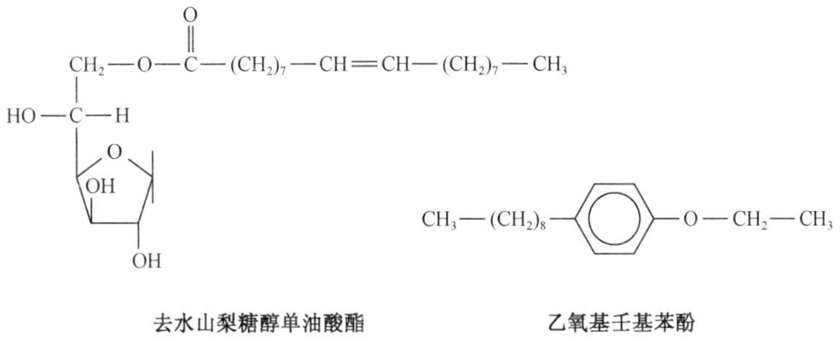

去水山梨糖醇单油酸酯   乙氧基壬基苯酚

图 3.9 浓缩液用表面活性剂

在含有黄胞胶的甲酸钠溶液中加入聚合物,可制备浓度为 15% 或更高的羟乙基纤维素(HEC)、疏水改性纤维素醚、疏水改性羟乙基纤维素(HEC)、甲基纤维素,羟丙基甲基纤维素和聚(环氧乙烷)的液化水基浓缩液,其中黄胞胶的作用是稳定剂(Burdick 和 Pullig,1993)。

制备浓缩液的顺序是:向水中先加入黄胞胶,然后是甲酸钠,最后是聚合物。

3.3 油基压裂液用稠化剂

3.3.1 有机稠化剂磷酸铝酯

磷酸二酯或磷酸二酯铝化物可用于合成柴油或原油的稠化剂(Gross,1987)。三元酸酯和五氧化二磷反应生成金属磷酸二酯,再合成聚磷酸盐,然后再与己醇反应生成磷酸二酯(Huddleston,1989)。

在有机溶剂中加入二元酸酯,连接到非水铝分子上,例如,以柴油为溶剂,利用异丙醇铝(图3.10)合成金属磷酸二酯,控制上述反应条件即可合成油基压裂液用稠化剂,并具有较好的耐温性和耐剪切性。所有的试剂中均不含水,也不会影响 pH 值。磷酸二酯的合成见图3.11,首先是磷酸三乙酯和五氧化二磷反应,最后再与己醇进行酯化反应。

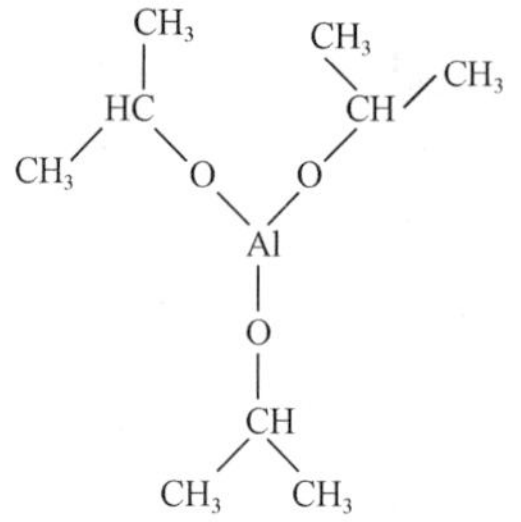

图3.10 异丙醇铝

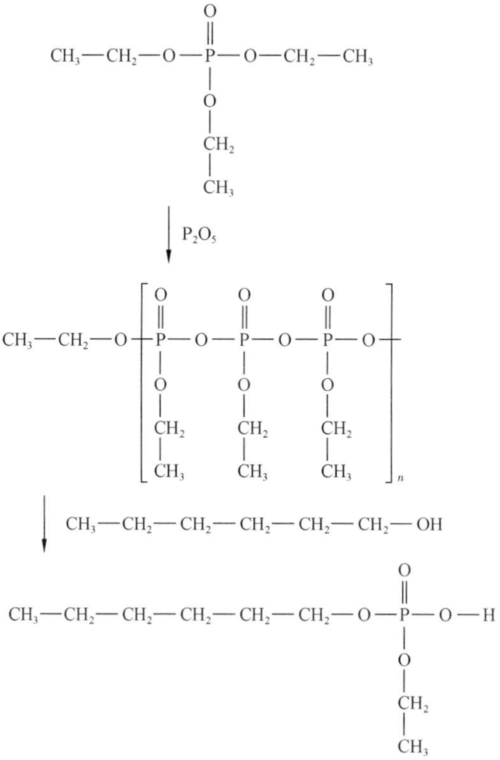

图3.11 磷酸二酯的合成

磷酸酯的增强剂是氨基化合物（Geib，2002）。2－乙基己酸三铝盐与脂肪酸一起可用作活化剂（Subramanian 等，2001b）。

合成油基烃稠化剂的另一种方法是用铁盐与正磷酸酯反应（Smith 和 Persinski，1995），而不用铝化合物。铁盐的优点如下：合成稠化剂时可含有高达 20% 的水，pH 值范围也比较宽。常用破胶剂也可用于该稠化剂形成的冻胶。

3.3.2 柴油增稠

N，*N*－二甲基丙烯酰胺和 *N*，*N*－二甲氨基丙烷甲基丙烯酰胺的共聚物、单羧酸和乙醇胺（Holtmyer 和 Hunt，1988）均可用于柴油或煤油稠化剂。这些化合物如图 3.12 所示。

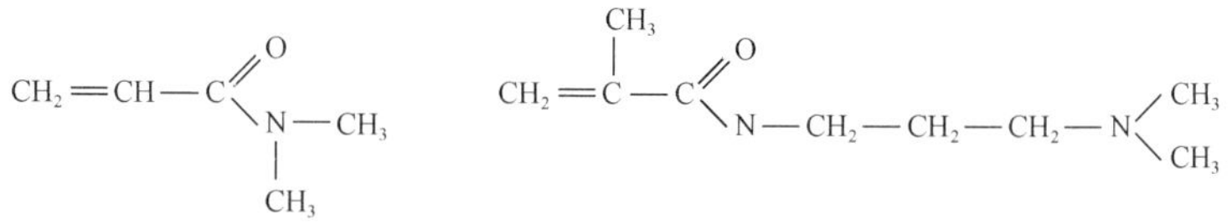

N，*N*-二甲基丙烯酰胺　　*N*，*N*-二甲氨基丙烷基甲基丙烯酰胺

图 3.12　柴油用稠化剂共聚物单体

3.4 黏弹性

在机械力作用下，黏弹性材料既表现出黏性又表现出弹性。在应力应变曲线中存在滞后现象，而且在恒定应变作用下，黏弹性材料存在应力松弛现象，应力逐渐减小。黏弹性材料还具有蠕变行为。

1867 年，James Clerk Maxwell 建立了黏弹性材料的简单模型（Maxwell，1867）。把理想化的黏性阻尼器和弹性弹簧连接在一起，用于模拟麦克斯韦液体或麦克斯韦材料，基本装置如图 3.13 所示。

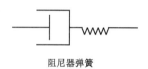

图 3.13　麦克斯韦模型

麦克斯韦模型既可用于解释弹性行为，也可用于解释黏性行为，一些基本问题也被重新定义（Rao 和 Rajagopal，2007），可以表示为：

$$\frac{\mathrm{d}\varepsilon_t}{\mathrm{d}t} = \frac{\mathrm{d}\varepsilon_\mathrm{d}}{\mathrm{d}t} + \frac{\mathrm{d}\varepsilon_\mathrm{s}}{\mathrm{d}t} = \frac{\sigma}{\eta} + \frac{1}{E}\frac{\mathrm{d}\sigma}{\mathrm{d}t} \tag{3.1}$$

总形变量 ε_t 随时间 t 的变化等于阻尼器形变量 ε_d 随时间的变化和弹簧形变量 ε_t 随时间的变化总和。方程式（3.1）右边包含简单的牛顿流变定律和理想化弹簧形变量的胡克定律，即：

$$\frac{\mathrm{d}\varepsilon_\mathrm{d}}{\mathrm{d}t} = \frac{\sigma}{\eta} \quad \varepsilon_\mathrm{s} = \frac{1}{E}\sigma$$

其中，η 是牛顿黏度，σ 是应力，E 是弹性模量。如果弹簧和阻尼器并联而不是串联的话，则采用 Kelvin－Voigt 模型，这应该是不成文的经验。

Rehage 和 Hoffmann 在 1988 年研究了表面活性剂黏弹性，阳离子表面活性剂水溶液的离子间相互作用力很强，性能与凝胶类似。

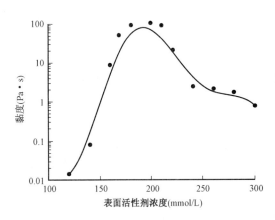

图 3.14　表面活性剂黏度与浓度的关系
（Yin 等,2009）

Yin 等利用流变学和动态光散射相结合的方法,研究阳离子表面活性剂黏弹性溶液形成的流变特性和微观结构的变化(Yin 等,2009),流变行为非常复杂。有关十二烷基三乙基溴化胺和十二烷基硫酸钠表面活性剂的研究表明:表面活性剂超过一定浓度时,柱状胶束增多,最终形成蠕虫状胶束;浓度继续增加,溶液呈现出带有麦克斯韦流体特征的线性黏弹性;当浓度更高时,线性胶束就会发展成分支结构。

当十二烷基三乙基溴化胺和十二烷基硫酸钠比例为 27∶73,在零剪切速率下,黏度与二者总浓度的对应关系如图 3.14 所示。

3.4.1　黏弹性表面活性剂稠化剂

大部分表面活性剂溶液中的球状胶束本身不会产生黏弹性,而是不同类型的胶束相互缠结所致。VES 水溶液之所以产生黏度归因于表面活性剂分子胶束,球形胶束本身没有黏度,然而,当胶束结构增长,例如杆状或蠕虫状胶束,彼此相互缠结致使液体黏度增加(Crews 和 Huang,2010)。

在表面活性剂溶液中,有的表面活性剂分子离开 VES 胶束结构进入溶液中,有的分子又从溶液中进入 VES 胶束结构,因此通常认为 VES 结构是动态的。

虽然 VES 溶液呈现出剪切变稀特性,但是重复施加高剪切应力,溶液仍然是稳定的。相比之下,典型的高分子稠化剂在受到高剪切应力作用后,高分子则会产生降解作用(Colaco 等,2007)。

VES 破胶指的是杆状或蠕虫状胶束转换成球状胶束,换句话说,就是具有黏性的长胶束到非黏性球状结构的重组(Crews 和 Huang,2010)。

3.4.2　剪切恢复增强剂

一些黏弹性表面活性剂在受到高剪切作用后恢复能力较差,该性能阻碍其在油田深井中的应用。剪切恢复增强剂能够缩短黏弹性表面活性剂剪切后的恢复时间。

剪切恢复增强剂是烷基化聚葡萄糖苷、聚葡萄糖胺或乙二醇乙醚丙烯酸酯的共聚物。葡萄糖胺含有环状葡萄糖,半缩醛基上的氢被烷基或芳基取代。葡萄糖苷和葡萄糖胺基本结构如图 3.15 所示。

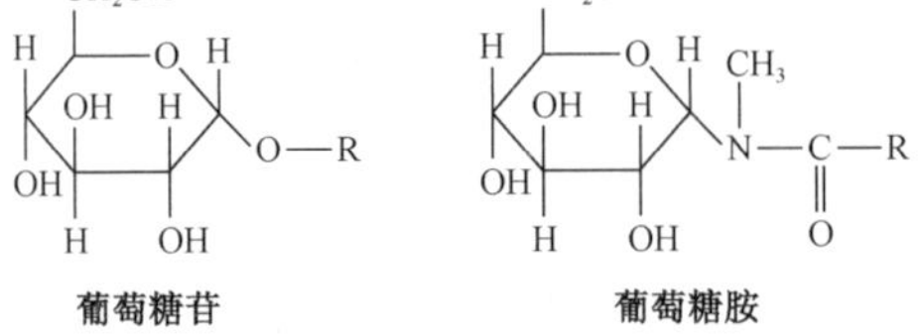

图 3.15　葡萄糖苷和葡萄糖胺结构(Colaco 等,2007)

黏弹性表面活性剂溶液是由适当的表面活性剂混配而成。当表面活性剂用量远超过临界浓度,再受到电解质的影响,分子聚集,形成胶束结构,胶束相互缠结再形成网络结构,即表现出黏弹性性能(Sullivan 等,2006)。

在表面活性剂溶液中加入某些试剂,即可形成黏弹性表面活性剂溶液。表面活性剂是长链季铵盐,如十六烷基三甲基溴化铵。

文献中的商品名称

商品名	描述	供应商
Alkaquat™ DMB-451	烷基二甲基苄基氯化铵(Lawrence 和 Warrender,2010)	Rhodia Canada Inc.
Benol®	石蜡油(Crews 和 Huang,2010)	Sonneborn Refined Products
Captivates® liquid	鱼明胶和阿拉伯树胶胶囊(Crews 和 Huang,2010)	ISP Hallcrest
Carnation®	石蜡油(Crews 和 Huang,2010)	Sonneborn Refined Products
ClearFRAC™	增产工作液(Couillet 和 Hughes,2008;Crews,2006;Crews 和 Huang,2010)	Schlumberger Technology Corp.
Diamond FRAQ™	VES 破胶剂(Crews 和 Huang,2010)	Baker Oil Tools
DiamondFRAQ™	VES 体系(Crews 和 Huang,2010)	Baker Oil Tools
Escaid® (Series)	矿物油(Crews 和 Huang,2010)	Crompton Corp.
Geltone® (Series)	有机黏土(Lawrence 和 Warrender,2010)	Halliburton Energy Services, Inc.
Hydrobrite® 200	石蜡油(Crews 和 Huang,2010)	Sonneborn Inc.
Isopar® (Series)	异烷烃溶剂(Crews 和 Huang,2010)	Exxon
Poly-S. RTM	聚合物胶囊(Crews,2006;Crews 和 Huang,2010)	Scotts Comp.
Rhodafac® LO-11A	磷酸酯(Lawrence 和 Warrender,2010)	Rhodia Inc. Corp.
Span® 20	失水山梨醇月桂酸酯(Crews 和 Huang,2010)	Uniqema
Span® 40	失水山梨醇单棕榈酸酯(Crews 和 Huang,2010)	Uniqema
Span® 65	失水山梨醇三硬脂酸酯(Crews 和 Huang,2010)	Uniqema
Span® 80	失水山梨醇单油酸酯(Crews 和 Huang,2010)	Uniqema
Span® 85	失水山梨醇三油酸酯(Crews 和 Huang,2010)	Uniqema
Tween® 20	失水山梨醇月桂酸酯(Crews 和 Huang,2010)	Uniqema
Tween® 21	失水山梨醇月桂酸酯(Crews 和 Huang,2010)	Uniqema
Tween® 40	失水山梨醇单棕酸酯(Crews 和 Huang,2010)	Uniqema
Tween® 60	失水山梨醇单硬脂酸酯(Crews 和 Huang,2010)	Uniqema
Tween® 61	失水山梨醇单硬脂酸酯(Crews 和 Huang,2010)	Uniqema
Tween® 65	失水山梨醇三硬脂酸酯(Crews 和 Huang,2010)	Uniqema
Tween® 81	失水山梨醇油酸酯(Crews 和 Huang,2010)	Uniqema
Tween® 85	失水山梨醇油酸酯(Crews 和 Huang,2010)	Uniqema
VES-STA 1	凝胶稳定剂(Crews 和 Huang,2010)	Baker Oil Tools
WS-44	乳化剂(Welton 等,2010)	Halliburton Energy Services, Inc.

参 考 文 献

Bahamdan, A., Daly, W. H., 2007. Hydrophobic guargumderivatives prepared by controlled graftingprocesses. Polym. Adv. Technol. 18(8), 652 – 659. http://dx.doi.org/10.1002/pat.874.

Baranet, S. E., Hodge, R. M., Kucera, C. H., 1987. Stabilized fracture fluid and crosslinker therefor. US Patent 4 686 052, assigned to Dowell Schlumberger Inc., Tulsa, OK, 11 August 1987. < http://www.freepatentsonline.com/4686052.html >.

Batelaan, J. G., van der Horts, P. M., 1994. Method of making amide modified carboxyl – containing polysaccharide and fatty amide – modified polysaccharide so obtainable. WO Patent 9 424 169, assigned to Akzo Nobel NV, 27 October 1994.

Bharat, P., 1990. Well treating fluids and additives therefor. EP Patent 372 469, 13 June 1990.

Brannon, H. D., 1988. Fracturing fluid slurry concentrate and method of use. EP Patent 280 341, 31 August 1988.

Burba, J. L. I., Strother, G. W., 1991. Mixed metal hydroxides for thickening water or hydrophilicfluids. US Patent 4 990 268, assigned to Dow Chemical Co., 5 February 1991.

Burba, J. L. I., Hoy, E. F., Read Jr., A. E., 1992. Adducts of clay and activated mixed metal oxides. WO Patent 9 218 238, assigned to Dow Chemical Co., 29 October 1992.

Burdick, C. L., Pullig, J. N., 1993. Sodium formate fluidized polymer suspensions process. USPatent5 228 908, assigned to Aqualon Co., 20 July 1993.

Colaco, A., Marchand, J. – P., Li, F., Dahanayake, M. S., 2007. Viscoelastic surfactant fluids havingenhanced shear recovery, rheology and stability performance. US Patent 7 279 446, assigned toRhodia Inc., Cranbury, NJ, Schlumberger Technology Corp., Sugarland, TX, 9 October 2007. < http://www.freepatentsonline.com/7279446.html >.

Cooke Jr., C. E., 2009. Method and materials for hydraulic fracturing of wells using a liquiddegradable thermoplastic polymer. US Patent 7 569 523, 4 August 2009. < http://www.freepatentsonline.com/7569523.html >.

Couillet, I., Hughes, T., 2008. Aqueous fracturing fluid. US Patent 7 427 583, assigned toSchlumberger Technology Corporation, Ridgefield, CT, 23 September 2008. < http://www.freepatentsonline.com/7427583.html >.

Courtois – Sambourg, J., Courtois, B., Heyraud, A., Colin – Morel, P., Rinaudo – Duhem, M., 1993. Polymer compounds of the glycuronic acid, method of preparation and utilization particularlyas gelifying, thickening, hydrating, stabilizing, chelating or flocculating means. WO Patent9 318 174, assigned to Picardie Univ., 16 September 1993.

Crews, J. B., 2006. Bacteria – based and enzyme – based mechanisms and products for viscosityreduction breaking of viscoelastic fluids. US Patent 7 052 901, assigned to Baker Hughes Inc., Houston, TX, 30 May 2006. < http://www.freepatentsonline.com/7052901.html >.

Crews, J. B., Huang, T., 2010. Use of oil – soluble surfactants as breaker enhancers for ves – gelledfluids. US Patent 7 696 135, assigned to Baker Hughes Inc., Houston, TX, 13 April 2010. < http://www.freepatentsonline.com/7696135.html >.

Crews, J. B., Huang, T., Gabrysch, A. D., Treadway, J. H., Willingham, J. R., Kelly, P. A., Wood, W. R., 2010. Methods and compositions for fracturing subterranean formations. US Patent 7 723 272, assigned to Baker Hughes Inc., Houston, TX, 25 May 2010. < http://www.freepatentsonline.com/7723272.html >.

Deem, C. K., Schmidt, D. D., Molner, R. A., 1991. Use ofMMH (mixed metal hydroxide)/propyleneglycol mud for minimization of formation damage in a horizontal well. In: ProceedingsVolume. No. 91 – 29. 4th Cade/Caodc Spring Drilling Conf., Calgary, Canada, 10 – 12 April 1991, Proc.

Doherty, D. H., Ferber, D. M., Marrelli, J. D., Vanderslice, R. W., Hassler, R. A., 1992. Geneticcontrol of acetylation and pyruvylation of xanthan based polysaccharide polymers. WO Patent9 219 753, assigned to Getty Scientific Dev Co., 12 November 1992.

Doolan, J. G., Cody, C. A., 1995. Pourable water dispersible thickening composition for aqueoussystems and a method of thickening said aqueous systems. US Patent 5 425 806, assigned toRheox Inc. ,20 June 1995.

Egraz,J. B. ,Grondin,H. ,Suau,J. M. ,1994. Acrylic copolymer partially or fully soluble in water,cured or not and its use(copolymere acrylique partiellement ou totalement hydrosoluble,reticule ou non et son utilisation). EP Patent 577 526,assigned to Coatex SA,5 January 1994.

Felixberger, J. ,1996. Mixed metal hydroxides(MMH)—an inorganic thickener for waterbaseddrilling muds(Mixed Metal Hydroxide(MMH)—Ein anorganisches Verdickungsmittelfür wasserbasierte Bohrspülungen). In:Proceedings Volume. DMGK Spring Conf. ,Celle,Germany,25 – 26 April 1996,pp. 339 – 351.

Geib,G. G. ,2002. Hydrocarbon gelling compositions useful in fracturing formations. US Patent6 342 468,assigned to Ethox Chemicals Llc. ,29 January 2002.

Gotlieb,K. F. ,Bleeker,I. P. ,van Doren,H. A. ,Heeres,A. ,1996. 2 – nitroalkyl ethers of nativeormodified starch, method for the preparation thereof,and ethers derived therefrom. EP Patent710 671,assigned to Coop Verkoop Prod. Aard De,8 May 1996.

Gregory,G. ,Shuell,D. ,Thompson Sr. ,J. E. ,1991. Overview of contemporary lfc(liquid fracconcentrate) fracture treatment systems and techniques. In:Proceedings Volume. No. 91 – 01. 4th Cade/caodc Spring Drilling Conf. ,Calgary,Canada,10 – 12 April 1991,Proc.

Gross,J. M. ,1987. Gelling organic liquids. EP Patent 225 661,assigned to Dowell SchlumbergerInc. ,16 June 1987.

Guo, D. R. , Gao, J. P. , Lu, K. H. , Sun, M. B. , Wang, W. , 1996. Study on the biodegradability of mudadditives. Drilling Fluid and Completion Fluid 13(1) ,10 – 12.

Harms,W. M. ,Norman,L. R. ,1988. Concentrated hydrophilic polymer suspensions. US Patent4 772 646,20 September 1988.

Harms,W. M. ,Watts,M. ,Venditto,J. ,Chisholm,P. ,1988. Diesel – based hpg(hydroxypropyl guar) concentrate is product of evolution. Pet. Eng. Int. 60(4) ,51 – 54.

Hodge, R. M. , 1997. Particle transport fluids thickened with acetylate free xanthanheteropolysaccharide biopolymer plus guar gum. US Patent 5 591 699,assigned to Du Pont DeNemours andand Co. ,7 January 1997.

Holtmyer,M. D. ,Hunt,C. V. ,1988. Method and composition for viscosifying hydrocarbons. USPatent 4 780 221,25 October 1988.

Holtmyer, M. D. , Hunt, C. V. , 1992. Crosslinkable cellulose derivatives. EP Patent 479 606, assignedto Halliburton Co. ,8 April 1992.

House, R. F. , Cowan, J. C. , 2001. Chitosan – containing well drilling and servicing fluids. US Patent6 258 755, assigned to Venture Innovations Inc. ,10 July 2001.

Huddleston,D. A. ,1989. Hydrocarbon geller andmethod formaking the same. US Patent 4 877 894,assigned to Nalco Chemical Co. ,31 October 1989.

Jafry,H. R. ,Pasquali,M. ,Barron,A. R. ,2011. Effect of functionalized nanomaterials on therheology of borate cross – linked guar gum. Ind. Eng. Chem. Res. 50(6) ,3259 – 3264. http://dx. doi. org/10. 1021/ie101836z.

Jones,T. G. J. ,Tustin,G. J. ,2010. Process of hydraulic fracturing using a viscoelastic wellbore fluid. US Patent 7 655 604,assigned to Schlumberger Technology Co. , Ridgefield, CT, 2 February2010. < http://www. freepatentsonline. com/7655604. html >.

Keilhofer,G. ,Plank,J. ,2000. Solids composition based on clay minerals and use thereof. US Patent6 025 303,assigned to Skw Trostberg AG,15 February 2000.

Kelly, P. A. , Gabrysch, A. D. , Horner, D. N. , 2007. Stabilizing crosslinked polymer guars andmodified guar derivatives. US Patent 7 195 065, assigned to Baker Hughes Inc. , Houston, TX, 27 March 2007. < http://www. freepatentsonline. com/7195065. html >.

Kotelnikov,V. S. ,Demochko,S. N. ,Melnik,M. P. ,Mikitchak,V. P. ,1992. Improving properties ofdrilling solution—by addition of ferrochrome – lignosulphonate and aqueous solution of cementand carboxymethyl cellulose. SU Patent

1 730 118, assigned to Ukr. Natural Gas Res. Inst. ,30 April 1992.

Kucera, C. H. , 1987. Fracturing of subterranean formations. US Patent 4 683 068, assigned toDowellSchlumberger Inc. ,Tulsa,OK,28 July 1987. < http://www. freepatentsonline. com/4683068. html >.

Lange, P. , Plank, J. ,1999. Mixed metal hydroxide(MMH) viscosifier for drilling fluids: Propertiesand mode of action (MixedMetal Hydroxide(MMH)—Eigenschaften und Wirkmechanismusals Verdickungsmittel in Bohrspülungen). Erdöl Erdgas Kohle 115(7 – 8) ,349 – 353.

Langlois, B. ,1998. Fluid comprising cellulose nanofibrils and its use for oil mining. WO Patent9 802 499,assigned to Rhone Poulenc Chimie,22 January 1998.

Langlois, B. , Guerin, G. , Senechal, A. , Cantiani, R. , Vincent, I. , Benchimol, J. , 1999. Fluidcomprising cellulose nanofibrils and its use for oil mining(fluide comprenant des nanofibrillesde cellulose et son application pour l'exploitation de gisements petroliers). EP Patent 912 653,assigned to Rhodia Chimie,6 May 1999.

Lawrence,S. , Warrender, N. , 2010. Crosslinking composition for fracturing fluids. USPatent 7 749 946, assigned to Sanjel Co. ,Calgary,Alberta,CA,6 July 2010. < http://www. freepatentsonline. com/7749946. html >.

Lemanczyk,Z. R. ,1992. The use of polymers in well stimulation: An overview of application, performance and economics. Oil Gas Europe Mag. 18(3) ,20 – 26.

Li,F. ,Dahanayake,M. ,Colaco,A. ,2010. Multicomponent viscoelastic surfactant fluid and methodof using as a fracturing fluid. US Patent 7 772 164, assigned to Rhodia, Inc. , Cranbury, NJ, 10August 2010. < http://www. freepatentsonline. com/7772164. html >.

Lukach,C. A. , Zapico, J. , 1994. Thermally stable hydroxyethylcellulose suspension. EP Patent619 340, assigned to Aqualon Co. ,12 October 1994.

Lundan, A. O. , Lahteenmaki, M. J. ,1996. Stable cmc(carboxymethyl cellulose) slurry. US Patent5 487 777,assigned to Metsa Serla Chemicals Oy,30 January 1996.

Lundan, A. O. , Anas, P. H. , Lahteenmaki, M. J. ,1993. Stable cmc(carboxymethyl cellulose) slurry. WO Patent 9 320 139,assigned to Metsa Serla Chemicals Oy,14 October 1993.

Maxwell,J. C. ,1867. On the dynamical theory of gases. Phil. Trans. R. Soc. Lond. 157(1) ,49 – 88.

McElfresh,P. M. , Williams, C. F. , 2007. Hydraulic fracturing using non – ionic surfactant gellingagent. US Patent 7 216 709, assigned to Akzo Nobel N. V. , Arnhem, NL, 15 May 2007. < http://www. freepatentsonline. com/7216709. html >.

Meyer, V. , Audibert – Hayet, A. , Gateau, J. P. , Durand, J. P. , Argillier, J. F. ,1999. Water – solublecopolymers containing silicon. GB Patent 2 327 946,assigned to Inst. Francais Du Petrole,10 February 1999.

Mondshine, T. C. , 1987. Crosslinked fracturing fluids. WO Patent 8 700 236, assigned to TexasUnited Chemical Corp. ,15 January 1987.

Mueller,H. ,Breuer,W. ,Herold,C. P. ,Kuhm,P. ,von Tapavicza,S. ,1997. Mineral additives forsetting and/or controlling the rheological properties and gel structure of aqueous liquid phasesand the use of such additives. US Patent 5 663 122,assigned to Henkel KG Auf Aktien,2September 1997.

Patel,B. B. ,1994. Fluid composition comprising a metal aluminate or a viscosity promoter anda magnesium compound and process using the composition. EP Patent 617 106,assigned to Phillips Petroleum Co. ,28 September 1994.

Pelissier,J. J. M. ,Biasini,S. ,1991. Biodegradable drilling mud(boue de forage biodegradable). FRPatent 2 649 988, 25 January 1991.

Putzig, D. E. ,2012. Zirconium – based cross – linking composition for use with high pH polymersolutions. US Patent 8 153 564, assigned to Dorf Ketal Speciality Catalysts, LLC, Stafford, TX, 10 April 2012. < http://www. freepatentsonline. com/8153564. html >.

Rangus,S. ,Shaw,D. B. ,Jenness,P. ,1993. Cellulose ether thickening compositions. WO Patent9 308 230,assigned to Laporte Industries Ltd. ,29 April 1993.

Rao, I. J., Rajagopal, K. R., 2007. On a new interpretation of the classical Maxwell model. Mech. Res. Commun. 34(7 − 8), 509 − 514. < http://www. sciencedirect. com/science/article/B6V48 − 4P6M5XY − 3/2/55d0b1078e2f1113c002c004729d5079 >.

Rehage, H., Hoffmann, H., 1988. Rheological properties of viscoelastic surfactant systems. J. Phys. Chem. 92(16), 4712 − 4719.

Robinson, F., 1996. Polymers useful as pH responsive thickeners and monomers therefor. WOPatent 9 610 602, assigned to Rhone Poulenc Inc., 11 April 1996.

Robinson, F., 1999. Polymers useful as pH responsive thickeners and monomers therefor. US Patent5 874 495, assigned to Rhodia, 23 February 1999.

Rummo, G. J., Startup, R., 1986. Bisalkyl bis(trialkanol amine)zirconates and use ofsame as thickening agents for aqueous polysaccharide solutions. US Patent4 578 488, assigned to Kay − Fries, Inc., Stony Point, NY, 25 March 1986. < http://www. freepatentsonline. com/4578488. html >.

Smith, K. W., Persinski, L. J., 1995. Hydrocarbon gels useful in formation fracturing. US Patent5 417 287, assigned to Clearwater Inc., 23 May 1995.

Subramanian, S., Islam, M., Burgazli, C. R., 2001a. Quaternary ammonium salts as thickeningagents for aqueous systems. WO Patent 0 118 147, assigned to Crompton Corp., 15 March2001.

Subramanian, S., Zhu, Y. P., Bunting, C. R., Stewart, R. E., 2001b. Gelling system for hydrocarbonfluids. WO Patent 0 109 482, assigned to Crompton Corp., 8 February 2001.

Sullivan, P., Christanti, Y., Couillet, I., Davies, S., Hughes, T., Wilson, A., 2006. Methodsfor controlling the fluid loss properties of viscoelastic surfactant based fluids. US Patent7 081 439, assigned to Schlumberger Technology Corporation, Sugar Land, TX, 25 July 2006. < http://www. freepatentsonline. com/7081439. html >.

Waehner, K., 1990. Experience with high temperature resistant water based drilling fluids(Erfahrungen beim Einsatz hochtemperatur − stabiler wasserbasischer Bohrspülung). ErdölErdgas Kohle 106(5), 200 − 201.

Welton, T. D., Todd, B. L., McMechan, D., 2010. Methods for effecting controlled break in pHdependent foamed fracturing fluid. US Patent 7 662 756, assigned to Halliburton EnergyServices, Inc., Duncan, OK, 16 February 2010. < http://www. freepatentsonline. com/7662756. html >.

Westland, J. A., Lenk, D. A., Penny, G. S., 1993. Rheological characteristics of reticulated bacterialcellulose as a performance additive to fracturing and drilling fluids. In: Proceedings Volume. SPE Oilfield Chemical International Symposium, NewOrleans, 3 − 5 March 1993, pp. 01 − 514.

Wilkerson, J. M. I., Verstrat, D. W., Barron, M. C., 1995. Associative monomers. US Patent 5 412 142, assigned to National Starch and Chemical Investment holding Corp., 2 May 1995.

Yeh, M. H., 1995a. Anionic sulfonated thickening compositions. EP Patent 632 057, assigned toRhone Poulenc Spec. Chem. C, 4 January 1995.

Yeh, M. H., 1995b. Compositions based on cationic polymers and anionic xanthan gum(compositions a base de polymeres cationiques et de gomme xanthane anionique). EP Patent654 482, assigned to Rhone Poulenc Spec. Chem. C, 24 May 1995.

Yin, H., Lin, Y., Huang, J., 2009. Microstructures and rheological dynamics of viscoelasticsolutions in a catanionic surfactant system. J. Colloid Interface Sci. 338(1) 177 − 183. < http://www. sciencedirect. com/science/article/B6WHR − 4WGK6P7 − 2/2/38daea0593cc7f3234 ba25febeefe035 >.

Zeilinger, S. C., Mayerhofer, M. J., Economides, M. J., 1991. A comparison of the fluid − lossproperties of borate − zirconate − crosslinked and noncrosslinked fracturing fluids. In: ProceedingsVolume. SPE East Reg Conference, Lexington, KY, 23 − 25 October 1991, pp. 201 − 209.

4 降阻剂

在油气井钻井、完井和储层改造过程中,通常要通过油管、连续油管等管柱泵注某种工作液,而这些工作液在管注中的湍流流动会消耗大量能量,因此需要增加泵注压力才能达到预期改造效果(Robb 等,2010)。

延迟交联作用能够降低泵注摩阻,有些特定添加剂也可降低管道阻力。在油井压裂改造中最先采用的降阻剂是瓜尔胶,目前已经成为最常用的做法。

在压裂液中加入相对少量的生物纤维素(0.60~1.8g/L)可提高液体的流变性能(Penny 等,1991),提高支撑剂悬浮能力,降低摩阻损失。

4.1 不配伍性

尽管四元表面活性剂与杀菌剂配伍,但是有些四元表面活性剂与阴离子降阻剂却不配伍,而这些降阻剂在井下施工时又非常有用。

有人认为两种带电分子会导致四元表面活性剂与降阻剂反应,并最终形成沉淀,两种添加剂不配伍。此外,某些杀菌剂如氧化剂能够降解特定的降阻剂(Bryant 等,2009)。

4.2 聚合物

为了降低泵注过程中的能量损失,可以在工作液中加入某种降阻聚合物。在高温井施工时可加入聚丙烯酰胺(PAM)、聚丙烯酸酯和共聚物,并且任何井施工都可以加入低浓度降阻剂(Lukocs 等,2007)。

降阻聚合物一般都是高分子聚合物,分子质量至少在 2.5MDa 以上,并且一般都是线性的,具有一定弹性。丙烯酰胺(AM)和丙烯酸的共聚物(Robb 等,2010)就是一种降阻聚合物。合成降阻聚合物的单体统计见表 4.1。

表 4.1 降阻剂聚合物单体(Robb 等,2010)

降阻剂聚合物单体	降阻剂聚合物单体
丙烯酰胺	丙烯酸
2-丙烯酰氨基-2-甲基丙烷磺酸	N,N-二甲基丙烯酰胺
乙烯基磺酸	N-乙烯基乙酰胺
N-乙烯基甲酰胺	亚甲基丁二酸
甲基丙烯酸	丙烯酸酯类
甲基丙烯酸酯	

降阻聚合物可以是酸或盐化合物。现场用水如果含有多价离子则与降阻聚合物相互作用，影响降阻剂效果，因此加入络合剂络合多价离子，可提高降阻剂性能。

为提高降阻聚合物性能，将络合剂加入到浓缩聚合物中比直接加入水中用量更低。据观察，将无机络合剂加入不相溶的油连续相中，可大大提高降阻剂性能。络合剂见表4.2。

表 4.2　络合剂（McMechan 等，2009；Robb 等，2010）

化合物	化合物
碳酸盐	柠檬酸
磷酸盐	葡萄糖酸
焦磷酸盐	葡萄糖庚酸
正磷酸盐	乙二胺四乙酸
次氮基三乙酸	

4.3　环境因素

考虑到环保问题，降阻聚合物的使用还存在一定问题（King 等，2007）。例如，之前用到的很多降阻聚合物都是油外相乳化聚合物，添加到工作液中以后发生乳液转化，释放出降阻聚合物。

虽然降阻剂在水基前置液和其他水基工作液中已经得到成功应用，但是降阻剂悬浮在油水乳化液中有毒，并且对环境有害。因此，必须改进降阻剂，使其成为无毒无害产品（King 等，2004）。

在工作液的后期处理过程中，油外相乳液中的碳氢化合物可能会带来环境问题，而在某些地区或会受到严格的环境法规限制（King 等，2007）。另外，油外相乳液中的碳氢化合物也会污染地下水资源，考虑到环境因素，水基降阻聚合物应该比其他聚合物类型更适合用作降阻剂。

在二甲氨基甲基丙烯酸酯上接枝氯化苄、丙烯酰乙氧基三甲基氯化铵稳定分散均聚物，然后再与丙烯酰胺反应可合成无毒环保型降阻剂（King 等，2004）。单体结构如图4.1所示。

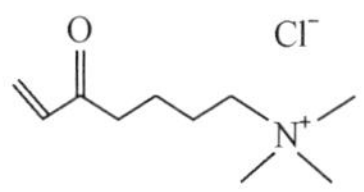

图 4.1　丙烯酰乙氧基三甲基氯化铵

4.4　二氧化碳泡沫压裂液

2-丙烯酰氨基-2-甲基-1-丙烷磺酸（AMPS）在二氧化碳泡沫压裂液中的降阻能力见表4.3，其中二氧化碳体积浓度为20%。

表 4.3 泡沫压裂液用 AMPS 基聚合物降阻剂

聚合物类型	AMPS(%)	降阻能力(%)
丙烯酰胺,痕量丙烯酸酯	0	0
15%~20% AMPS、丙烯酰胺、痕量丙烯酸酯	15~20	13
阳离子丙烯酰胺	0	0
<10% AMPS、丙烯酰胺、痕量丙烯酸酯	10	0
20%~25% AMPS、丙烯酰胺、痕量丙烯酸酯	20~25	22
40% AMPS、丙烯酰胺、痕量丙烯酸酯	40	39

聚合物乳液:

用 AM 和阴离子单体合成的共聚物通常都是降黏组分,阴离子单体有丙烯酰氨基丙烷磺酸、丙烯酸、甲基丙烯酸、单丙烯乙氧基磷酸盐或相应的碱金属盐(Parnell 等,2012)。

通过添加相对大量的表面活性剂,可增加聚合物的亲水性和分散性,从而增加液体的稳定性。此外,即使在低能量条件下,增加表面活性剂量也可增加添加剂的反转率。因此,少量聚合物的加入可提高溶液降阻性能。

然而,反相表面活性剂可能与乳化剂或乳液产生不利反应,在使用前就破坏了乳化剂或乳液。因此聚合物乳液中通常加入少于 5% 的反相表面活性剂。然而,含有少量反相表面活性剂的聚合物乳液不能降低摩阻,因为聚合物乳液既不能完全转化也不抗盐或抗酸。

这种配方是逐步加入两种可溶的表面活性剂组合形成的(Parnell,2012)。溶剂最好是萜烯,如 d-萜二烯、松油精、l-萜二烯、d,l-萜二烯、月桂烯和 α-蒎烯。表面活性剂是乙氧基甘油脂、乙氧基山梨醇酯、乙氧基烷基酚类、乙氧基醇、蓖麻油乙氧基化物、椰油胺乙氧基化物或山梨醇单油酸盐。表面活性剂的亲水亲油平衡值应该大于 7。

4.5 油外相共聚物乳液

采用乳液聚合可制备油外相共聚物,然后制得降阻共聚物。乳液聚合技术因引发温度、引发剂类型和加量、单体和搅拌速度的不同而不同。可用于合成油外相共聚物乳液的组分见表 4.4。

表 4.4 乳液聚合用化合物(Chatterji 等,2006)

组分	用量(%)
石蜡/环烷类有机溶剂	21.1732
妥尔油脂肪酸二乙醇胺	1.1209
聚氧乙烯失水山梨醇油酸酯	0.0722
失水山梨醇油酸酯	0.3014
丙烯酰胺	22.2248
4-甲氧基苯酚	0.0303
氯化铵	1.6191

续表

组分	用量(%)
丙烯酸	4.3343
乙二胺四乙酸	0.0237
叔丁基氢过氧化物	0.0023
焦亚硫酸钠	0.2936
2,2′-偶氮(2-脒基丙烷)盐酸盐	0.0311
乙氧基化 C_{12}—C_{16} 醇	1.3700
水	43.1737

用不同浓度的丙烯酰胺和丙烯酸合成的降阻剂共聚物进行降阻实验,结果见表4.5。

表 4.5　降阻实验数据(Chatterji 等,2006)

时间(min)	丙烯酰胺/丙烯酸			
	70/30	85/15	87.5/12.5	90/10
4	65.9	66.3	62.2	57.2
9	61.0	56.1	54.3	50.2
14	55.2	49.8	50.3	45.2
19	50.0	45.8	45.7	41.3
最大时间	69.7	71.1	70.7	69.7

4.6　带有弱不稳定链接的聚丙烯酰胺

在特定条件下,聚合物主链中含有弱链接的水溶性聚合物可以降解,比如温度、pH 值或还原剂,降解效果很好(Kot 等,2012)。根据需要,按照预定方案可对该聚合物进行可控降解,降解后变成小分子物质,溶液黏度降低,快速溶于水中,而且不易吸附在储层表面。此外,施工时不再需要加入氧化破胶剂,也不需要清除已沉积的聚合物,可节约施工时间。

事实证明,使用含有氧化还原电子对的双官能团还原剂有利于引发自由基聚合,将可降解基团连接到乙烯基聚合物的主链上。其中双官能团指的是可降解基团和氧化金属离子。

最令人期待的是可温度降解,但对水解稳定的偶氮基团。核磁共振光谱和差示扫描量热法可以证实在 PAM 主链上存在偶氮基团。凝胶渗透色谱法可用于描述含有温度敏感性偶氮基团的 PAM 的降解性能,还可以证明聚合物主链上连有多个不稳定链接。实验表明聚合物主链上含有偶氮的 PAM 与纯 PAM 降阻性能相近,然而,一旦温度升高,含有偶氮的 PAM 就会失去其降阻性能,有时这也是一种优势(Kot 等,2012)。

文献中的商品名称

商品名	描述	供应商
Alkamuls® SMO	失水山梨醇油酸酯(Chatterji 等,2006)	Rhodia HPCII
Amadol® 511	妥尔油脂肪酸二乙醇胺(Chatterji 等,2006)	Akzo Nobel Surface Chemistry
FR™(系列)	降阻剂(Robb 等,2010)	Halliburton Energy Services,Inc.
LPA® -210	有机溶剂(Chatterji 等,2006)	Sasol North America,Inc.
Surfonic®	乳化剂,乙氧基化 C_{12} 醇(Chatterji 等,2006)	Huntsman Performance Products
Tween® 81	失水山梨醇油酸酯(Chatterji 等,2006)	Uniqema

参 考 文 献

Bryant,J. E. ,McMechan,D. E. ,McCabe,M. A. ,Wilson,J. M. ,King,K. L. ,2009. Treatment fluidshaving biocide and friction reducing properties and associated methods. US Patent Application20090229827, 17 September 2009. < http://www. freepatentsonline. com/20090229827. html >.

Chatterji,J. ,King,K. L. ,McMechan,D. E. ,2006. Subterranean treatment fluids,friction reducingcopolymers,and associated methods. US Patent 7 004 254,assigned to Halliburton EnergyServices,Inc. ,Duncan,OK,28 February 2006. < http://www. freepatentsonline. com/7004254. html >.

Harris,P. C. ,Heath,S. J. ,2006. Friction reducers for fluids comprising carbon dioxide andmethods of using friction reducers in fluids comprising carbon dioxide. US Patent7 117 943,assigned to Halliburton Energy Services,Inc. ,Duncan,OK,10 October 2006. < http://www. freepatentsonline. com/7117943. html >.

King,K. L. ,McMechan,D. E. ,Chatterji,J. ,2004. Methods,aqueous well treating fluids and frictionreducers therefor. US Patent 6 784 141,assigned to Halliburton Energy Services,Inc. ,Duncan,OK,31 August 2004. < http://www. freepatentsonline. com/6784141. html >.

King,K. L. ,McMechan,D. E. ,Chatterji,J. ,2007. Water - based polymers for use as friction reducersin aqueous treatment fluids. US Patent 7 271 134,assigned to Halliburton Energy Services,Inc. ,Duncan,OK,18 September 2007. < http://www. freepatentsonline. com/7271134. html >.

Kot,E. ,Saini,R. ,Norman,L. R. ,Bismarck,A. ,2012. Novel drag - reducing agents for fracturingtreatments based on polyacrylamide containing weak labile links in the polymer backbone(includes supplementary experimental section). SPE J. 17(3),924 - 930. < http://www. onepetro. org/mslib/app/Preview. do? paperNumber = SPE - 141257 - PA&societyCode = SPE >.

Lukocs,B. ,Mesher,S. ,Wilson,T. P. J. ,Garza,T. ,Mueller,W. ,Zamora,F. ,Gatlin,L. W. ,2007. Non - volatile phosphorus hydrocarbon gelling agent. US Patent Application 20070173413,assigned to Clearwater International,LLC. ,26 July 2007. < http://www. freepatentsonline. com/20070173413. html >.

McMechan,D. E. ,Hanes Jr. ,R. E. ,Robb,I. D. ,Welton,T. D. ,King,K. L. ,King,B. J. ,Chatterji,J. ,2009. Friction reducer performance by complexing multivalent ions in water. US Patent7 579 302,assigned to Halliburton Energy Services,Inc. ,Duncan,OK,25 August 2009. < http://www. freepatentsonline. com/7579302. html >.

Parnell,E. ,Sanner,T. ,Holtmyer,M. ,Philpot,D. ,Zelenev,A. ,Gilzow,G. ,Champagne,L. ,Sifferman,T. ,2012. Drag - reducing copolymer compositions. US Patent Application20120035085,assigned to Cesi Chemical,Inc. ,Marlow,OK,9 February 2012. < http://www. freepatentsonline. com/20120035085. html >.

Penny,G. S. ,Stephens,R. S. ,Winslow,A. R. ,1991. Method of supporting fractures in geologicformations and hydraulic fluid composition for same. US Patent 5 009 797,assigned toWeyerhaeuser Co. ,23 April 1991.

Robb,I. D. ,Welton,T. D. ,Bryant,J. ,Carter,M. L. ,2010. Friction reducer performance in watercontaining multivalent ions. US Patent 7 846 878,assigned to Halliburton Energy Services,Inc. ,Duncan,OK,7 December 2010. < http://www. freepatentsonline. com/7846878. html >.

5 液体滤失添加剂

已有研究者论述了压裂液滤失的基本问题(Harris,1988;Rimassa 等,2009)。流体滤失添加剂也被称为降滤失剂。当流体进入多孔地层,就会发生流体滤失。这与钻井完井液,压裂液和水泥浆都有关系。

流体滤失的程度依赖于孔隙度及相应的地层渗透率,可能达到约 10t/h。由于在某些情况相下,用于石油技术的流体相当昂贵,因此人们不能容忍大量的流体滤失。当然,防止流体滤失也有环境的原因。

5.1 液体滤失剂的作用机理

在某种程度上,降低滤失就是想办法堵塞岩石孔隙。基本原理见表 5.1。

表 5.1 滤失控制的机理

颗粒类型	描述
大颗粒	悬浮颗粒堵塞孔隙,形成滤饼降低渗透率
小颗粒	大分子在多孔储层边界层中形成凝胶
化学浆液	向地层中挤入树脂,不可逆的;适合于较大的洞穴

5.1.1 液体滤失的测量

因为许多用于评价的参数无法获得,流体滤失的过程很复杂,并且难以模拟。因此,一般使用经验模型,其中一些比较符合实际。某些测试方法也会影响到数据的质量。已有研究回顾了一些模型,并提出了其他的模型以提供更好的适合可用的数据(Clark,2010)。

静态滤失测量为比较压裂液材料或理解黏性流体浸入、滤饼形成和滤饼侵蚀的复杂机制提供了不充分的结果(Vitthal 和 McGowen,1996b);且动态滤失的研究还不足以形成适当的试验方法,从而导致错误或相互矛盾的结果。

将大规模高温高压模拟装置的实验结果与实验室数据相比,可以发现初滤失值有显著差异(Lord 等,1995)。

为了获得水泥固井过程中某些阶段的流体滤失行为信息,特别是当水泥浆候凝时,可以用活塞式静态滤失实验得到可靠结果。

实验结果表明,液体滤失可以分成两个明显的阶段(Vitthal 和 McGowen,1996a):

(1)早期高滤失阶段,在适当的滤饼形成之前,液体穿过地层表面(初滤失);

(2)所有液体滤失都是通过滤饼控制的阶段。

5.1.2 大颗粒的作用

早在 1995 年,Chin 的专著就给出了关于大颗粒侵入地层的机理。

预防流体滤失的基本原理之一如图 5.1 所示。流体中含有悬浮颗粒,这些颗粒自井眼径向流入地层孔隙。多孔地层就像一个筛子,将这些颗粒捕获,并在表面附近聚集成为滤饼。

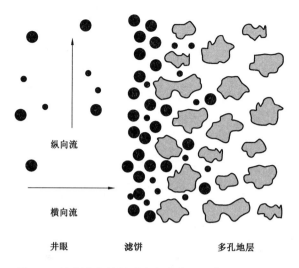

图 5.1 钻井液中的悬浮液在多孔地层表面形成滤饼

作用于悬浮物的流体动力决定了滤饼的形成速率,因此也就决定了流体的滤失率。已有文献使用简单的模型,预测了在滤饼表面,滤失速度和剪切应力之间的幂律关系(Jiao 和 Sharma,1994)。

该模型表明,在小颗粒滤失过程中,形成的滤饼是不均匀的。当悬浮液中没有足够小的颗粒可以沉积时,会形成一个平衡的滤饼厚度。作为时间的函数,可以从模型计算出滤饼厚度。

对于给定流变性和流速的悬浮液,存在滤饼临界渗透率,在该渗透率条件下,不会形成滤饼。通过选择合适的悬浮液流速和滤饼渗透率,平衡滤饼厚度是可以精确控制的。

沿着裂缝滤失到地层深部的压裂液或用于降低滤失量的材料,很容易伤害高渗裂缝区。高渗透地层压裂的几个特点是,短的裂缝长度,相对常规压裂往往不成比例的裂缝宽度,和正的压后表皮效应。这是裂缝表面伤害的结果(Aggour 和 Economides,1996)。如果该压裂液浸入量小,伤害程度是次要的。

所以如果压裂液滤失量较小,没有正表皮效应,即使严重的渗透率伤害都是可以容忍的。压裂设计的首要任务应该是获得最大裂缝导流能力。在高渗透地层压裂中,需要使用含降滤失剂和破胶剂的高浓度聚合物交联压裂液。

降滤失的材料也会伤害支撑带的导流能力。在裂缝端部的高剪切速率可能会阻止外部滤饼的形成,从而增加在高渗透地层的初滤失量,所以防止伤害性的添加剂是必要的。酶降解的液体降滤失剂是较好的选择。表 5.2 总结了一些适用于水力压裂液的降滤失剂。

表 5.2 水力压裂液用降滤失剂

化学药品	参考文献
碳酸钙和木质素磺酸盐①	Johnson 和 Smejkal(1993),Johnson(1996)
天然淀粉	Elbel 等(1995),Navarrete 和 Mitchell(1995),Navarrete 等(1996)
羧甲基淀粉	Elbel 等(1995),Navarrete 和 Mitchell(1995),Navarrete 等(1996)
羟丙基淀粉②	Elbel 等(1995),Navarrete 等(1996),Navarrete 和 Mitchell(1995)

续表

化学药品	参考文献
羧甲基纤维素(HEC)和交联瓜尔胶③	Cawiezel 等(1999)
颗粒状淀粉和云母微粒	Cawiezel 等(1999)

① 可以加入 Wellan 或黄胞胶聚合物以保证碳酸钙和木质素磺酸盐形成悬浮液。
② 协同效应,见文本。
③ 渗透率500mD。

由于未破胶聚合物冻胶残余在裂缝内部,瓜尔胶类的聚合物所造的人工裂缝的导流能力可能降低。这种残渣可能会伤害支撑带的渗透率,致使裂缝导流能力较低,并减少裂缝的有效长度。需要从冻胶伤害过程的两个重要方面进行评估,即滤饼厚度和裂缝内浓缩聚合物冻胶的屈服应力(Xu 等,2011)。

滤失过程中生成的滤饼厚度是聚合物浓度和滤失量的函数。滤饼厚度和滤失量呈线性关系。对于瓜尔胶聚合物压裂液,冻胶浓度的因素是个常数。由于滤失产生的浓缩聚合物滤饼,表现为具有屈服应力的 Herschel–Bulkley 流变特征。

Herschel–Bulkley 流体属于广义的非牛顿流体。在这里,流体所经历的应变与相应的应力关系是相当复杂的,这种关系是非线性的,可以用稠度系数 K,流动行为指数 n 和剪切屈服应力 τ_0 来表征。这种流体模型是1926年建立的(Herschel 和 Bulkley,1926)。

屈服应力是一个临界参数,表示冻胶是否可以从裂缝中去除。聚合物滤饼的屈服应力与聚合物及破胶剂的浓度紧密相关(Xu 等,2011)。

5.2 化学添加剂

5.2.1 淀粉和云母颗粒

一种液体降滤失剂是由颗粒淀粉和云母微粒组成的(Cawiezel,1996)。云母属于层状硅酸盐。表5.3 中列出了某些云母类矿物。

表5.3 云母矿物

名称	化学式
黑云母	$K(Mg,Fe)_3(AlSi_3)O_{10}(OH)_2$
白云母	$KAl_2(AlSi_3)O_{10}(OH)_2$
金云母	$KMg_3(AlSi_3)O_{10}(OH)_2$
锂云母	$K(Li,Al)_{2-3}(AlSi_3)O_{10}(OH)_2$
珍珠云母	$CaAl_2(Al_2Si_2)O_{10}(OH)_2$
海绿石	$(K,Na)(Al,Mg,Fe)_2(Si,Al)_4O_{10}(OH)_2$

地层压裂的方法是打井穿过地层,然后在足以压开地层的泵注排量和压力下,由井筒注入地层含有大量控制滤失添加剂的压裂液,连通地层。

5.2.2 解聚淀粉

部分解聚淀粉可以降低液体滤失,其黏度要小于相应的没有部分解聚的淀粉衍生物的黏度(Dobson 和 Mondshine,1997)。

5.2.3 可控制降解的液体滤失添加剂

天然淀粉和改性淀粉加上酶可以作为压裂液降滤失剂（Williamson 等，1991b）。如前所述，酶可以降解淀粉的 α 键，而没法降解作为稠化剂的瓜尔胶及其衍生物的 β 键。

天然淀粉和改性淀粉应用的推荐比例范围为 3:7 到 7:3，最佳为 1:1，混合物以干粉状态从地面加入到井筒中。建议优先考虑使用羧甲基和羟丙基的衍生物。天然淀粉来源于土豆、玉米、小麦和大豆，最好的是玉米淀粉。

混合包括两个或两个以上的改性淀粉，与天然和改性淀粉混合物相同。用表面活性剂包覆淀粉，如聚氧乙烯失水山梨醇单油酸酯，丁醇或乙氧基壬基苯酚，有助于淀粉在压裂液中的分散。

威廉姆森等在1991年描述了一种流体滤失添加剂，该添加剂具有较低的初滤失量和滤失速度，在压裂液进入地层时，可以快速形成低渗透滤饼，从而有助于达到所需的裂缝几何形状。施工结束后，该滤失添加剂很容易降解。该添加剂具有较宽的粒度分布，可以有效处理较宽地层孔隙度范围，并且易于在压裂液中分散。

该滤失添加剂是改性淀粉混合物，混合了一个或以上的改性淀粉和一个或更多的天然淀粉。这些混合物对于流体注入并形成裂缝比天然淀粉更有效。添加剂的降解是受控制的，是由氧化反应或地层存在的细菌攻击而形成可溶性的产物。可以通过加入过硫酸盐和过氧化物类氧化剂加速氧化过程。

优选的过氧化物是过氧化钙和过氧化镁（Giffin，2012）。过硫酸盐破胶剂一般为过硫酸铵，过硫酸钠和过硫酸钾（Card 等，1999；Tulissi 等，2012）。

液体滤失添加剂的降解：

研究人员给出了一种由天然淀粉（玉米淀粉）和化学改性淀粉（羧甲基和羟丙基淀粉）加上酶混合制成的压裂液降滤失添加剂（Williamson 和 Allenson，1989；Williamson，1991b）。酶可以降解淀粉的 α 键，而不会降解作为稠化剂的瓜尔胶及其衍生物的 β 键。

可以用表面活性剂包覆淀粉颗粒，如失水山梨醇油酸酯，乙氧基丁醇或乙氧基壬基苯酚，有助于淀粉在压裂液中的分散。改性淀粉或改性淀粉与天然淀粉的混合物具有较宽的粒径分布范围，比天然淀粉更有助于降低压裂液在裂缝中的滤失（Williamson 等，1991a）。这种淀粉可以用氧化或生物降解。

5.2.4 琥珀酰聚糖

琥珀酰聚糖是一种生物聚合物，具有理想的滤失控制性能（Lau，1994a，1994b）。包括：易于配制、清洁、剪切变稀、黏度在转变温度（T_m）以下对温度不敏感、并且转变温度（T_m）在较宽的温度范围内可以调节。琥珀酰聚糖溶液依靠黏度降低液体的滤失。不会形成对地层具有较大伤害的不易除去的滤饼。

基于这些认识，琥珀酰聚糖作为降滤失剂，已经在海上超过 100 口油井的砾石充填作业中应用。基于实验室测量的流变性能和现场经验的计算结果表明，琥珀酰聚糖对于 HEC 是无效的。使用适当的琥珀酰聚糖降滤失剂，可以使液体滤失从 40bbl/h 降低至仅有几桶每小时。大多数井在处理后的产量不受影响。

可以用内部酸破胶剂降解琥珀酰聚糖（Bouts 等，1997）。如果延迟作用的内部破胶剂起作用,由于黏度滤失控制剂未完全返排造成的地层伤害可以降低到最小。岩心流动实验表明,琥珀酰聚糖与内部盐酸破胶剂相结合形成降滤失体系,可以持续控制流动中的液体滤失,并延迟破胶,而伤害较低。为了描述琥珀酰聚糖/盐酸体系的延迟破胶过程,建立了基于键断裂率的模型。

利用该模型,可以在不同地层温度、琥珀酰聚糖转变温度和酸浓度条件,预测聚合物流变学特征随时间的变化。该模型可用来确定现场要求的琥珀酰聚糖与破胶剂优化配方,如液体滤失控制所需的时间间隔,可抵挡的过平衡压力,以及所需要的盐水密度。

5.2.5 硬葡聚糖

粒径分级的碳酸钙颗粒,硬葡聚糖非离子多糖和改性淀粉的混合物也可用于控制液体的滤失(Johnson,1996)。碳酸钙颗粒具有较宽的粒径范围非常重要,较宽的粒径范围有助于阻止液体滤失或滤液进入地层。由于生物高聚物和淀粉的作用,滤饼颗粒不会侵入井筒,在去除滤饼时也就不会出现高压峰值。

液体的流变学特征常用于保持所希望的原始地层渗透率。这些应用包括钻井、压裂和完井作业中的滤失控制,如砾石充填或修井作业。

5.2.6 聚原酸酯

脂肪族聚酯因水解而化学降解。水解的过程可以用酸和碱催化。水解过程中,断链时形成羟基基团,这样可以进一步加快水解速度。这种机制被艺术地称为链式反应作用,而且认为生成的聚酯基团大部分被降解。

在酯类,聚原酸酯和脂肪族聚酯之中,优先考虑聚丙交酯。聚丙交酯可以用乳酸的缩合反应或用更常见的环丙交酯单体开环聚合反应合成获得。

水解降解的速度相当慢。在压裂作业中,在裂缝内铺置支撑剂之前,材料不应开始降解。此外,缓慢的降解有助于支撑剂铺置过程中压裂液滤失的控制(Todd等,2006)。表5.4中给出了一种压裂液的配方。

表5.4 压裂液配方(Todd等,2006)

化合物	商品名[①]	比例(%)
水		
氯化钾		1
破乳剂	LO - SURF 300	0.05
聚酯		0.15
瓜尔胶		0.2
缓冲剂(CH_3COOH)	BA - 20	0.005
强碱	MO - 67	0.1
硼交联剂	CL - 28M	0.05
破胶剂($NaClO_3$)	VICON NF	0.1
杀菌剂(2,2 - 二溴 - 3 - 次氮基丙酰胺)	BE - 3S	0.001
杀菌剂(2 - 溴 - 2 - 硝基 - 1,3 - 丙二醇)	BE - 6	0.001
压裂砂		50

① 哈里伯顿能源服务公司。

5.2.7 聚(羟基乙酸)

低分子量羟基乙酸或含有羟基酸、羧酸、羟基羧酸根等其他化合物的混合物的缩合产物,可以作为降滤失剂(Bellis 和 McBride,1987)。已经有关于聚合物生产方法的描述。

反应产品颗粒粒径为 0.1~1500μm。缩合产物可以在水力压裂过程中,用作可水解的水冻胶压裂液的降滤材料。

羟基乙酸缩合物在地层条件下水解成为羟基乙酸,这打破了水冻胶的链式反应,并最终确保地层渗透率恢复,而无需单独添加破胶剂(Cantu 等,1990,1989;Casad 等,1991)。

5.2.8 多酚

研究人员描述了用于油基钻井液的亲油机质的多酚材料(Cowan 等,1988)。该添加剂是用多酚氧化酶与一种或更多的磷脂合成得到的。磷脂是从植物油中获得的甘油磷脂,最好的植物油是商业卵磷脂。

腐殖酸、木质素磺酸、木质素、酚醛固化物、单宁或这些多酚材料的氧化、磺化或磺甲基化衍生物,可以作为多酚的原料。

还可以使用分级碳酸钙和改性木质素磺酸盐作为降滤失剂(Johnson 和 Smejkal,1993)。可以选用任意一种触变聚合物,如 Wellan 或黄胞胶聚合物,作为碳酸钙和木质素磺酸盐的悬浮剂。

重要的是,碳酸钙粒子应有较宽的粒径范围,以防止滤失或滤液进入地层。此外,木质素磺酸盐必须聚合以便在有效范围内降低其水溶解度。在井筒壁面形成滤饼需要改性的木质素磺酸盐。

因为由于改性木质素磺酸盐的作用,滤饼粒子并未侵入井筒,在清除滤饼过程中不会出现高压峰值,表明对地层和井筒表面的损伤较小。这种添加剂可以用于压裂液、完井液和修井液。

试验表明,基于磺化单宁酚醛树脂的液体降滤失剂可以在高温高压条件下有效控制液体的滤失,并表现出良好的耐盐和耐酸性能(Huang,1996)。

5.2.9 作为转向材料的邻苯二甲酰亚胺

如图 5.2 所示,邻苯二甲酰亚胺是一种转向材料或降滤失剂,用于包括酸在内的水基处理液的转向,可以进入地层逐步降低渗透率(Dill,1987)。

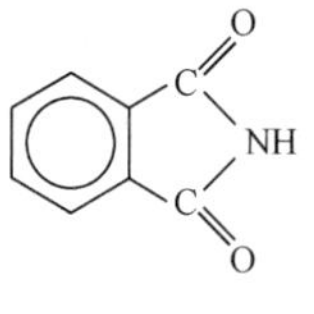

图 5.2 邻苯二甲酰亚胺

该添加剂同样可以降低水基或烃基液体向地层的滤失,如在压裂施工中。材料的性能取决于材料的颗粒大小。

邻苯二甲酰亚胺可承受较高的地层温度,同时,能够通过溶于地层产出液或在高温下升华,很容易从地层中排除出去。该材料要与其他降低地层渗透率和增加地层渗透率的两种材料都配伍。邻苯二甲酰亚胺颗粒由井筒进入地层,封堵裂缝、孔隙、喉道和岩穴,从而堵塞部分地层,起到降滤失的作用。

5.2.10 黏弹性添加剂

在表面活性剂溶液中产生黏弹性的试剂是盐,如氯化铵、氯化钾、水杨酸钠、异氰酸酯钠。此外,氯仿等非离子有机分子,有助于产生黏弹性。表面活性剂的电解质浓度也是形成黏弹性的重要参数(Sullivan 等,2006)。黏弹性表面活性剂(VES)水基冻胶压裂液已经用于水力压裂施工中。

然而,同样的致使 VES 液体对地层伤害小的特性,造成滤失进入地层基质的量较高,从而降低了 VES 压裂施工中的液体效率。因此,在高渗地层中进行 VES 压裂时,使用降滤失剂是非常重要的(Huang 和 Crews,2009)。

可以添加矿物油提高这种液体的滤失性能,在环境温度下,这种矿物油在环境温度下的黏度大于 20mPa·s。这种矿物油最初以油滴状分散在流体的内相,即不连续相内。在冻胶化后,将矿物油添加到液体中。

在下面可重复实验中,用 AkzoNobel 公司提供的牛脂酰氨基丙基胺氧化物作为 VES 表面活性剂(Podwysock,2004)。已经证明,在含有 3%氯化钾和用 6%VES 在 66℃进行稠化的水溶液中,与不加矿物油相比,加入 2%矿物油对液体黏度有降低影响。这种特征与其他观察结果形成对照,因为大量的碳氢化合物和矿物油往往会抑制 VES 冻胶的形成或使 VES 冻胶破胶(Huang and Crews,2009)。

如模拟实验所示,可以改善滤失效果。图 5.3 给出了液体滤失作为测试时间函数的曲线。试验是在 0.7MPa,66℃条件下,用 400mD 人造岩心做的。

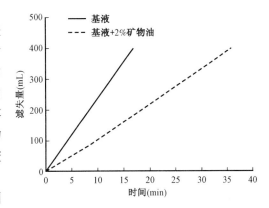

图 5.3 液体滤失量与时间的关系
(Huang 和 Crews,2009)

已经发现,在 VES 水基冻胶中加入氧化镁或氢氧化钙,可以改善这些盐水的流体滤失性能(Huang 等,2009)。

这些液体滤失控制剂的缓慢溶解性能是非常重要的,这样,这些滤失剂就能够很容易从地层中排出,并且对地层很少或完全没有伤害。

在 VES 水基冻胶体系中加入这些添加剂,可以限制和减少在压裂或压裂充填作业中 VES 溶液滤失进地层孔隙中的量,从而降低了由于 VES 溶液进入地层孔隙造成的伤害。

此外,在储层渗透率范围内,这种降滤失剂并不显著控制流体的滤失速度。因此,2000mD 的地层中的滤失速度可以与 100mD 的地层中的滤失速度相比较。这种特征增大了 VES 体系的储层渗透率应用范围。

可以确定,该液体滤失剂的作用与 VES 微胶团相关。当 VES 液体滤失进入储层时,微胶团在地层表面形成一层黏性胶束,滤失控制剂颗粒也会堆积在一起,从而降低了 VES 液体的滤失速度。

颗粒堵塞储层的孔隙并不是控制滤失的机理。用纳米级液体滤失控制剂进行试验,明确显示在1mD或更高渗透率的储层中,并未形成桥塞或堵塞,而在表面上仍然存在一个黏性胶束层。因此,流体滤失剂的颗粒大小不是控制滤失速率的关键或主要因素(Huang等,2009)。

降滤失剂的降解

研究人员给出了一种由天然淀粉(玉米淀粉)和化学改性淀粉(羧甲基和羟丙基淀粉)加上酶混合制成的压裂液降滤失添加剂(Williamson和Allenson,1989;Williamson,1991b)。酶可以降解淀粉的α键,而不会降解作为稠化剂的瓜尔胶及其衍生物的β键。

可以用表面活性剂包覆淀粉颗粒,如失水山梨醇油酸酯,乙氧基丁醇或乙氧基壬基苯酚,有助于淀粉在压裂液中的分散。改性淀粉或改性淀粉与天然淀粉的混合物具有较宽的粒径分布范围,比天然淀粉更有助于降低压裂液在裂缝中的滤失(Williamson等,1991a)。这种淀粉可以用氧化或生物降解。

文献中的商品名称

商品名	描述	供应商
Britolo® 35USP	高黏矿物油(Huang和Crews,2009)	Sonneborn Refined Products
ClearFRAC™	增产措施工作液(Huang和Crews,2009;Huang等,2009)	Schlumberger Technology Corp
Diamond FRAQ™	VES破胶剂(Huang和Crews,2009)	Baker Oil Tools
Gloria®	高黏矿物油(Huang和Crews,2009)	Sonneborn Refined Products
Hydrobrite® 200	石蜡油(Huang和Crews,2009)	Sonneborn Inc.
Kaydol® oil	矿物油(Huang和Crews,2009)	Witco Corp.
Performance® 225N	基础油(Huang和Crews,2009)	ConocoPhillips
SurFRAQ™ VES	牛脂酰氨基丙基胺氧化物(Huang和Crews,2009;Huang等,2009)	Baker Oil Tools

参 考 文 献

Aggour, T. M., Economides, M. J., 1996. Impact of fluid selection on high – permeability fracturing. In: Proceedings-Volume. vol. 2. SPEEurope PetroleumConference, Milan, Italy, 22 – 24 October1996, pp. 281 – 287.

Bellis, H. E., McBride, E. F., 1987. Composition and method for temporarily reducing thepermeability of subterranean formations. EP Patent 228 196, assigned to Du Pont De Nemours & Co., 8 July 1987.

Bouts, M. N., Trompert, R. A., Samuel, A. J., 1997. Time delayed and low – impairment fluid – losscontrol using a succinoglycan biopolymer with an internal acid breaker. SPE J. 2(4), 417 – 426.

Cantu, L. A., McBride, E. F., Osborne, M. W., 1989. Formation fracturing process. US Patent4 848 467, assigned to Conoco Inc. and Du Pont De Nemours & Co., 18 July 1989.

Cantu, L. A., McBride, E. F., Osborne, M., 1990. Well treatment process. EP Patent 404 489, assignedto Conoco Inc. and Du Pont De Nemours & Co., 27 December 1990.

Card, R. J., Nimerick, K. H., Maberry, L. J., McConnell, S. B., Nelson, E. B., 1999. On – theflycontrol of delayed borate – crosslinking of fracturing fluids. US Patent 5 877 127, assigned to Schlumberger Technology Corporation, Sugar Land, TX, 2 March 1999. < http://www. freepatentsonline. com/5877127. html >.

Casad, B. M., Clark, C. R., Cantu, L. A., Cords, D. P., McBride, E. F., 1991. Process for the preparationof fluid loss additive and gel breaker. US Patent 4 986 355, assigned to Conoco Inc., 22 January1991.

Cawiezel, K. E., Navarrete, R., Constien, V., 1996. Fluid loss control. GB Patent 2 291 906, assignedto Sofitech NV, 7 February 1996.

Cawiezel, K. E. , Navarrete, R. C. , Constien, V. G. ,1999. Fluid loss control. US Patent 5 948 733, assigned to Dowell Schlumberger Inc. ,7 September 1999.

Chin, W. C. ,1995. Formation Invasion: With Applications to Measurement – While – Drilling, TimeLapse Analysis, and Formation Damage. Gulf Publishing Co. , Houston.

Clark, P. E. ,2010. Analysis of fluid loss data II: models for dynamic fluid loss. J. Petrol. Sci. Eng. 70(3 – 4),191 – 197. http://dx. doi. org/10. 1016/j. petrol. 2009. 11. 010.

Cowan, J. C. , Granquist, V. M. , House, R. F. ,1988. Organophilic polyphenolic acid adducts. USPatent 4 737 295, assigned to Venture Chemicals Inc. ,12 April 1988.

Dill, W. R. ,1987. Diverting material and method of use for well treatment. CA Patent 1 217 320,3 February 1987.

Dobson, J. W. , Mondshine, K. B. ,1997. Method of reducing fluid loss of well drilling and servicingfluids. EP Patent 758 011, assigned to Texas United Chem. Co. Llc. ,12 February 1997.

Elbel, J. L. , Navarrete, R. C. , Poe Jr. , B. D. ,1995. Production effects of fluid loss in fracturing highpermeabilityformations. In: Proceedings Volume. SPE Europe Formation Damage ControlConference, The Hague, Netherlands,ˊ5 – 16 May 1995, pp. 201 – 211.

Giffin, W. J. ,2012. Compositions and processes for fracturing subterranean formations. US Patent8 293 687, assigned to Titan Global Oil Services Inc. , Bloomfield Hills, MI,23 October 2012. <http://www. freepatentsonline. com/8293687. html>.

Harris, P. ,1988. Fracturing – fluid additives. J. Petrol. Technol. 40(10). http://dx. doi. org/10. 2118/17112 – PA.

Herschel, W. H. , Bulkley, R. ,1926. Konsistenzmessungen von Gummi – Benzollösungen. Kolloid – Zeitschrift 39(4),291 – 300. http://dx. doi. org/10. 1007/BF01432034.

Huang, N. ,1996. Synthesis of fluid loss additive of sulfonate tannic – phenolic resin. Oil DrillingProd. Technol. 18(2),39 – 42,106 – 107.

Huang, T. , Crews, J. B. ,2009. Use of mineral oils to reduce fluid loss for viscoelastic surfactantgelled fluids. US Patent 7 615 517, assigned to Baker Hughes Inc. , Houston, TX, 10 November2009. <http://www. freepatentsonline. com/7615517. html>.

Huang, T. , Crews, J. B. , Treadway Jr. , J. H. ,2009. Fluid loss control agents for viscoelastic surfactantfluids. US Patent 7 550 413, assigned to Baker Hughes Inc. , Houston, TX,23 June 2009. <http://www. freepatentsonline. com/7550413. html>.

Jiao, D. , Sharma, M. M. ,1994. Mechanism of cake buildup in crossflow filtration of colloidalsuspensions. J. Colloid Interface Sci. 162(2),454 – 462.

Johnson, M. ,1996. Fluid systems for controlling fluid losses during hydrocarbon recoveryoperations. EP Patent 691 454, assigned to Baker Hughes Inc. ,10 January 1996.

Johnson, M. H. , Smejkal, K. D. ,1993. Fluid system for controlling fluid losses during hydrocarbonrecovery operations. US Patent 5 228 524, assigned to Baker Hughes Inc. ,20 July 1993.

Lau, H. C. ,1994a. Laboratory development and field testing of succinoglycan as a fluid – loss – controlfluid. SPE Drill. Completion 9(4),221 – 226.

Lau, H. C. ,1994b. Laboratory development and field testing of succinoglycan as fluid – loss – controlfluid. SPE Peer Approved Paper.

Lord, D. L. , Vinod, P. S. , Shah, S. , Bishop, M. L. ,1995. An investigation of fluid leakoff phenomenaemploying a high – pressure simulator. In: Proceedings Volume. Annual SPE TechnicalConference, Dallas,22 – 25 October 1995, pp. 465 – 474.

Navarrete, R. C. , Mitchell, J. P. ,1995. Fluid – loss control for high – permeability rocks in hydraulicfracturing under realistic shear conditions. In: Proceedings Volume. SPE Production and Operations Symposium, Oklahoma, City,2 – 4 April 1995, pp. 579 – 591.

Navarrete, R. C., Brown, J. E., Marcinew, R. P., 1996. Application of new bridging technology andparticulate chemistry for fluid – loss control during fracturing highly permeable formations. In: Proceedings Volume. vol. 2. SPE Europe Petroleum Conference, Milan, Italy, 22 – 24 October1996, pp. 321 – 325.

Podwysocki, M., 2004. Akzo nobel surfactants. Technical Bulletin SC05 – 0707, Akzo Nobel SurfaceChemistry LLC, 525 W. Van Buren Street Chicago, IL 60607 – 3823.

Rimassa, S., Howard, P., Blow, K., 2009. Optimizing fracturing fluids from flowback water. In: Proceedings of SPE Tight Gas Completions Conference. No. 125336 – MS. SPE Tight GasCompletions Conference, 15 – 17 June 2009, San Antonio, Texas, USA, Society of PetroleumEngineers, Dallas, Texas, pp. 1 – 8. http://dx. doi. org/10. 2118/125336 – MS.

Sullivan, P., Christanti, Y., Couillet, I., Davies, S., Hughes, T., Wilson, A., 2006. Methodsfor controlling the fluid loss properties of viscoelastic surfactant based fluids. US Patent7 081 439, assigned to Schlumberger Technology Corporation, Sugar Land, TX, 25 July 2006. < http://www. freepatentsonline. com/7081439. html >.

Todd, B. L., Slabaugh, B. F., Munoz Jr., T., Parker, M. A., 2006. Fluid loss control additives for usein fracturing subterranean formations. US Patent 7 096 947, assigned to Halliburton EnergyServices, Inc., Duncan, OK, 29 August 2006. http://www. freepatentsonline. com/7096947. html.

Tulissi, M. G., Luk, S., Vaughan, J., Browne, D. J., Dusterhoft, D., 2012. Fracturing method andapparatus utilizing gelled isolation fluid. US Patent 8 141 638, assigned to TricanWellServicesLtd., Calgary, CA, 27 March 2012. < http://www. freepatentsonline. com/8141638. html >.

Vitthal, S., McGowen, J., 1996a. Fracturing fluid leakoff under dynamic conditions: Part 2: Effect of shear rate, permeability, and pressure. In: Proceedings of SPE Annual Technical Conference and Exhibition. No. 36493 – MS. SPE Annual Technical Conference and Exhibition, 6 – 9 October 1996, Denver, Colorado, Society of Petroleum Engineers, pp. 821 – 835. http://dx. doi. org/10. 2118/36493 – MS.

Vitthal, S., McGowen, J. M., 1996b. Fracturing fluid leakoff under dynamic conditions: Part. 2: Effect of shear rate, permeability, and pressure. In: ProceedingsVolume. Annual SPE TechnicalConference, Denver, 6 – 9 October 1996, pp. 821 – 835.

Williamson, C. D., Allenson, S. J., 1989. A new nondamaging particulate fluid – loss additive. In: Proceedings Volume. SPE International Symposium on Oilfield Chemistry, Houston, 8 – 10February 1989, pp. 147 – 158.

Williamson, C. D., Allenson, S. J., Gabel, R. K., 1991a. Additive andmethod for temporarily reducingpermeability of subterranean formations. US Patent 4 997 581, assigned toNalco Chemical Co., 5 March 1991.

Williamson, C. D., Allenson, S. J., Gabel, R. K., Huddleston, D. A., 1991b. Enzymatically degradablefluid loss additive. US Patent 5 032 297, assigned to Nalco Chemical Co., 16 July 1991.

Xu, B., Hill, A. D., Zhu, D., Wang, L., 2011. Experimental evaluation of guar – fracture – fluid filtercakebehavior. SPE Prod. Oper. 26(4), 381 – 387.

6 乳 化 剂

乳液在石油行业中发挥重要作用。这其中主要用于钻井液和压裂液等。其实在钻井液等方面的应用过程中,所指的乳液也不是单纯物理学上的乳液,而是如油基钻井液或水基钻井液等,这本质上并不是乳液。在本章中,主要介绍并总结一些基本的油田用乳液。

油田用乳液有时是基于它的动力学稳定性划分的(Kokal 和 Wingrove,2000;Kokal,2006)。主要类型如下:

(1)松散乳液。一般在几分钟内分层,分离出的水为自由水;
(2)介质乳液。乳液体系通常在 10min 内分层。
(3)稳定乳液。乳液体系稳定,体系通常可维持几小时、几天,甚至几周时间,并且很难完全分离。

乳剂将两种液体分散成小尺寸连续相即成为乳液。当乳液的分散相液滴大于 $0.1\mu m$ 时即为大液滴乳液(Kokal,2006)

单纯从热力学角度分析,乳液是一种不稳定的体系。这是因为乳液内部会自发地减少两相的界面面积和界面能,成为液—液分离系统。(Kokal 和 Wingrove,2000)。

第二种乳液被称为微乳液。这种乳液是两种不混溶液相以极低的界面能自发地混合在一起。微乳液液滴的尺寸很小,一般小于 10nm,具有极强的热力学稳定性。微乳液的稳定性与上面描述的大液滴乳液有本质上的不同。Santanna 于 2012 年综述了微乳液体系在石油工业中的应用(Santanna 等,2012)。

6.1 水包油乳液

水包油乳液已经较多应用于压裂作业。为了使乳液具有足够高的黏度,压裂液体系中的油相浓度必须足够高,一般可达到95%。在如此高的油相浓度下,体系通常难以稳定,液体中的油相通常形成球形油滴并逐渐离散出来。这样压裂液往往摩阻很高。为了提高乳液稳定性,需添加两相增强剂,如无机盐和可溶盐,从而稳定水包油乳液的稳定性能(Kiel,1973)。

在乳液中加入具有表面活性的增稠剂,可以促进乳化剂形成水包油乳液。例如含长链氨基聚乙烯羧酸与氢氧化钠反应,可促进形成稳定的水包油乳状液。乳液体系具有良好的耐高温性能,可用于高温深井压裂(Kiel,1973)。

6.2 反向乳液

反相乳液的连续相是油性液体,少量不溶于油的液相分散在其中成为不连续相。事实上,反相乳液是油包水乳液。

转相剂可以像钻井液中的钻屑一些样具有良好的悬浮性能。这样,可以较容易地加入液体体系中。众所周知,通过改变体系的 pH 值或添加表面活性剂改变液体表面活性可以使乳液转相。这种情况下,表面活性剂的亲水和亲油性能的转变,使体系内的连续相和不连续相发生转变(Taylor 等,2009)。

例如,如果在井筒内残留一定量的反相乳液,这部分乳液可以将原有的常规乳液清除出井筒,有利于降低井底伤害。反相乳液组合物可用于有机相和凝胶中。例如,利用癸烷膦酸单酯和 Fe^{3+} 活化剂可以使柴油形成胶凝体系(Taylor 等,2009)。

经常使用聚合物增加水溶液的黏度。聚合物如果亲水性能不强,可以通过形成乳液提高聚合物的水溶性(Jones 和 Wentzler,2008)。

6.3 水—水乳液

当两种或更多的可溶性聚合物溶液混合时,某些情况下会形成液—液混合相或形成各相分离状态。特别是,当两种具有不同热力学性能的聚合物水溶液相遇时,往往容易分层形成两相。这种乳液是水—水乳液或双水相液体系统(Sullivan 等,2010)。

在食品工业中,用这种液体配制聚合物溶液,用以模拟脂肪球的性能。在医药行业,水—水乳液体系用于蛋白质与酶的分离。

高分子聚合物双水相系统在油田也被广泛应用。这些混合物可以用来产生较低黏度的预水化混合物,这使得聚合物可以与水快速溶解分散,有利于现场大规模连续作业施工。

可以利用这种乳液原理,形成水相分离状态,解决瓜尔胶、羟丙基纤维素(HPC)等天然大分子聚合物的快速溶解问题。将干燥的瓜尔胶粉和干燥的 HPC 经搅拌混合成两相分离状态,然后继续搅拌成混合相。只需轻轻搅拌富含瓜尔胶和 HPC 的混合相体系就可以实现两相分离。通过改变离子强度,两相聚合物体系可以形成热力学弹性凝胶体系(Sullivan 等,2010)。这种措施可以用于区域隔离。

6.4 油包水包油乳液

油包水包油乳液可用于提高石油采收率或作为润滑流体传动液。相比于常规油包水乳液,这种乳剂的主要特性是可提高液体的剪切稳定性和剪切变稀。

油包水包油乳液的制备过程,先制备水包油乳液,然后将乳液再次分散在第二种油相中(Varadaraj,2010)。第二种油相中可以含有稳定剂,即微米至亚微米大小的固体颗粒,如环烷酸和沥青质。

6.5 微乳液

微乳液是一种热力学稳定流体。其与动力学稳定体系的乳液具有本质区别,常规乳液长时间放置会分离成油水两相,而微乳液不会。微乳液的粒径为 10~300nm。因为分散液滴的颗粒尺寸较小,宏观上微乳液会表现为透明或半透明状态。

微乳液的粒径可以用动态光散射和中子散射测定。微乳液具有超低界面张力。

油包水型微乳液已用于油田开发中。另外一种含有防腐剂的油—酒精微乳液,可以用做油田防冻剂(Yang 和 Jovancicevic,2009)。

微乳液可以广泛用于各种油溶性石油化学品,包括防腐剂、沥青抑制剂、阻垢剂等。从而降低有机溶剂的消耗量。微乳液提高了相关油田化学剂在液体体系中的溶解性,使得油田现场施工过程中,可以更容易地向其中添加更多的化学添加剂(Yang 和 Jovancicevic,2009)。

微乳液的分离及破坏可以通过各种方式,如化学品或温度的变化。然而破坏微乳液状态最简单的方式是稀释。

表6.1中列举了含缓蚀剂的微乳液。按照表6.1的加量形成的微乳液可以很容易在水相中稀释。如果增加甲苯的量,则微乳液可以由碳氢溶剂所稀释。当然这只是稀释微乳液的一部分例子,还有其他的办法(Yang 和 Jovancicevic,2009)。

表6.1 含缓蚀剂的微乳液(Yang 和 Jovancicevic,2009)

成分	含量(%)
甲苯	2
油酸咪唑啉(缓蚀剂)	4
油酸(腐蚀抑制剂)	4
十二烷基苯磺酸	2
乙醇胺	2
丁醇	20
水	60

6.6 固相稳定乳液

乳液的部分亲油性能可以使用不溶解的固体颗粒(Bragg,2000)。已有文献评估了带有固体颗粒的三相乳液的稳定性能(Menon 和 Wasan,1988)。此外,还评估了油泥乳液层的油滤失现象。因此可以用一些半经验方法,评估油的滤失性能。

固体颗粒可以是现有的或外加的。非地层内固体颗粒包括黏土、石英、长石、石膏、煤尘、沥青质和聚合物。但是一些固体颗粒中含有少量的离子化合物。通常情况下,颗粒具有复合不规则形状特点(Bragg,2000)。

固体颗粒应该有一些亲油特性,有助于在油外相乳液中的分散。这个性能可以确保固体颗粒稳定的悬浮于油外相中而不与内部水相接触。

固体颗粒的亲油和亲水性能是物质本身所固有的,也可以通过化学剂处理改变亲油亲水性能。

例如,亲油气相二氧化硅法,采用二氧化硅粉末或硅石粉,将二氧化硅粉末进行表面疏水处理,并最终形成直径为 $10\sim20nm$ 的小球,形成的固体粉末可以有效提高乳液的稳定性。固相颗粒的加量范围为 $0.5\sim20g/L$。

图6.1描述了带有固体颗粒乳剂的乳液在75/s剪切速率下水含量与黏度变化曲线。

乳液中的油可以与磺化剂进行预处理,以提高固体在乳液中的稳定性(Varadaraj 等,2004):

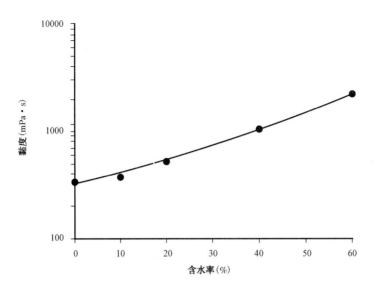

图 6.1　黏度与含水率变化关系图（Varadaraj 等,2004）

制备 6.1：原油与固体颗粒的磺化反应。将 12g 原油和固体颗粒（0.06g 的 2－甲基苄基牛油及蒙皂石和 0.12g 沥青）混合,50℃下搅拌 72h。然后每 100 份的原油中加入 3 份的浓硫酸,50℃继续搅拌 24h。

根据制备 6.1 的结果,磺化原油可以用由乙烯和丙烯制备的马来酸酐共聚物合成（Varadaraj 等,2007）。

此外,在石油化学中已经用光化学法替代磺化法进行处理（Varadaraj 等,2007）。因此,在进行光化学处理前,将膨润土凝胶与原油混合。在染料敏化光化处理过程中,原油首先与罗丹明 B 混合。罗丹明 B 是一种增加石油转化成氧化产品的光化学转换量子效率的红色染料。在表 6.2 中,给出了用各种方法预处理的原油成分的差异。

表 6.2　对原油预处理的各种方法的有效性

方式	含量(%)			
	饱和状态	未饱和	NSO[①]	沥青质
未处理原油	35.4	39.8	15.4	9.4
干燥光化学处理法	34.2	26.6	26.6	12.7
光化学处理法	31.1	20.5	30.7	17.9
热空气氧化	34.2	19.3	33.6	13.0
生物氧化	32.4	39.8	18.4	9.4

① NSO 化合物为芳香族化合物与氧、氮、硫等缩聚反应物。

6.7　生物处理的乳液

另一种预处理方法是通过微生物来提高油包水乳液的稳定性。这个过程使用石油降解微生物对原油进行生物处理（Varadaraj 等,2007）。

制备 6.2：将含有微生物的原油放入生物反应器中，然后加入 10～100 倍过量的水。石油降解微生物可以采用接种后的细菌。接种是一种培养微生物的方法。接种的微生物浓度可以通过菌落单位数量来确定。石油降解微生物可以用于油田污水处理。

可以通过营养物质为微生物提供食物来源。营养物质主要是富含氮和磷的物质。生物反应器中使用净化空气或氧气，温度范围为 20～70℃。

经过生物处理后，原油可以在生物反应液相中形成稳定的油包水乳液。然而，优化的方法是将形成的乳液与生物处理的生物油和水相混合，这样有利于提高乳化油的水稳定性。

以下几个步骤可以用于原油的生物处理，形成稳定的油包水乳液（Varadaraj 等，2007）：

（1）油中的一些脂肪成分可以通过氧化、极性酮或酸处理成脂肪族链。有机硫化合物容易氧化形成相应的亚砜类化合物。含氧化合物比脂肪类表面活性剂具有更好的活性，因而有助于提高油包水乳液的稳定性。

（2）如果环烷酸与二价阳离子如钙盐同时存在，则细菌可能会将这些盐脱羧形成环烷烃或低碳环烷酸以及相应的金属氧化物。这些成分可以提高油包水乳液的稳定性。

（3）在生物乳化原油过程中，生物反应得到的水相也发生较大变化。在生物反应完成后，水相中的表面活性剂相继分散，微生物反应会产生鼠李糖脂和死的微生物细胞。这些反应产物之间会起到协同作用，有助于提高油包水乳液的稳定性。因此，生物反应的水相可以用来形成油包水乳剂，进一步提高乳液稳定性。

文献中的商品名称

商品名	描述	供应商
Aerosil™（Series）	气相二氧化硅（Bragg，2000）	Evonik Goldschmidt GmbH
CAB-O-SIL™（Series）	气相二氧化硅（Bragg，2000）	Cabot. Corp.
Tegopren™（Series）	硅氧烷乳化剂（Jones and Wentzler，2008）	Evonik Goldschmidt GmbH
Tomadol®	脂肪胺（Jones and Wentzler，2008）	Tomah Products, Inc

参 考 文 献

Bragg, J. R., 2000. Oil recovery method using an emulsion. US Patent 6 068 054, assigned to ExxonProduction Research Co., 30 May 2000.

Jones, T. A., Wentzler, T., 2008. Polymer hydration method using microemulsions. USPatent 7 407 915, assigned to Baker Hughes Inc., Houston, TX, 5 August 2008. <http://www.freepatentsonline.com/7407915.html>.

Kiel, O. M., 1973. Method of fracturing subterranean formations using oil-in-water emulsions. US Patent 3 710 865, assigned to Esso Production Research Co., 16 January 1973. <http://www.freepatentsonline.com/3710865.html>.

Kokal, S. L., 2006. Crude oil emulsions. In: Fanchi, J. R. (Ed.), Petroleum Engineering Handbook. vol. I. Society of Petroleum Engineers, Richardson, TX, pp. 533-570 (Chapter 12). <http://balikpapan.spe.org/images/balikpapan/articles/51/Crude%20Oil%20Emulsions%20-%20PEH%20-%20Chapter%2012%20-%20compressed%20version.pdf>.

Kokal, S. L., Wingrove, M., 2000. Emulsion separation index: from laboratory to field case studies. In: Proceedings Volume. No. 63165-MS. SPE Annual Technical Conference and Exhibition, 1-4 October 2000, Society of Petroleum Engineers, Dallas, Texas.

Menon, V. B., Wasan, D. T., 1988. Characterization of oil – water interfaces containing finely dividedsolids with applications to the coalescence of water – in – oil emulsions: a review. ColloidsSurf. 29(1), 7 – 27. < http://www.sciencedirect.com/science/article/ B6W92 – 44CRG9F – 3/2/082bfad854dc70aa811b9ca060b31a1e >.

Santanna, V. C., de Castro Dantas, T. N., DantasNeto, A. A., 2012. The use of microemulsionsystems in oil industry. In: Najjar, R. (Ed.), Microemulsions—An Introduction toProperties and Applications. InTech Europe, Rijeka, Croatia, pp. 161 – 174 (Chapter 8). < http://www.intechopen.com/books/microemulsions – an – introduction – to – properties – and – applications/the – use – of – microemulsion – systems – in – oil – industry >.

Sullivan, P. F., Tustin, G. J., Christanti, Y., Kubala, G., Drochon, B., Hughes, T. L., 2010. Aqueous two – phase emulsion gel systems for zone isolation. US Patent 7 703 527, assigned to Schlumberger Technology Corporation, Sugar Land, TX, 27 April 2010. < http://www.freepatentsonline.com/7703527.html >.

Taylor, R. S., Funkhouser, G. P., Dusterhoft, R. G., 2009. Gelled invert emulsion compositionscomprising polyvalent metal salts of an organophosphonic acid ester or an organophosphinicacid and methods of use and manufacture. US Patent 7 534 745, assigned to HalliburtonEnergy Services, Inc., Duncan, OK, 19 May 2009. < http://www.freepatentsonline.com/7534745.html >.

Varadaraj, R., 2010. Oil – in – water – in – oil emulsion. US Patent 7 652 074, assigned to Exxon Mobil Upstream Research Co., Houston, TX, 26 January 2010. < http://www.freepatentsonline.com/7652074.html >.

Varadaraj, R., Bragg, J. R., Dobson, M. K., Peiffer, D. G., Huang, J. S., Siano, D. B., Brons, C. H., Elspass, C. W., 2004. Solids – stabilized water – in – oil emulsion and method for using same. USPatent 6 734 144, assigned to ExxonMobil Upstream Research Co., Houston, TX, 11 May 2004. < http://www.freepatentsonline.com/6734144.html >.

Varadaraj, R., Bragg, J. R., Peiffer, D. G., Elspass, C. W., 2007. Stability enhanced water – in – oilemulsion and method for using same. US Patent 7 186 673, assigned to ExxonMobilUpstream Research Co., Houston, TX, 6 March 2007. < http://www.freepatentsonline.com/7186673.html >.

Yang, J., Jovancicevic, V., 2009. Microemulsion containing oil field chemicals useful for oil andgas field applications. US Patent 7 615 516, assigned to Baker Hughes Inc., Houston, TX, 10November 2009. < http://www.freepatentsonline.com/7615516.html >.

7 破 乳 剂

部分压裂液体系中需要加入破乳剂,可以将黏性乳化液破乳为低黏状态,利于压后返排(Crowell,1984)。然而,在一些体系中,破乳剂也可用于降解聚合物。不含聚合物的乳化液流动性好,可快速返排。

将大量油内相分散到少量水外相中,可形成高黏度乳化液,但同时必须加入表面活性剂才能使其稳定。乳化液要转化为油包水状态或转化成各组分相,才能降低乳化液黏度,促进压后返排。

只有消除表面活性剂的稳定因素,如表面活性剂吸附在储层表面或者加入破乳剂,乳化液才能破乳。由于阳离子表面活性剂具有砂岩亲和力,因此容易吸附在储层表面。另外,如果要使用破乳剂,需优化破乳剂和表面活性剂种类及用量,确保乳化液正好在压后破乳。即使优选了合适的破乳剂和表面活性剂,也很难确定破乳准确时间,可能会过早破乳或者延误很长时间才能破乳(Graham 等,1976)。

聚合物乳化压裂液的高黏度归因于聚合物和乳化液两种因素,即使压裂液性能优良,也很难从地层中排出,因此要考虑这两种降黏机制。只有破乳作用和聚合物降解作用同时发生,才能使聚合物乳化液转化成低黏液体。破胶剂最好在压裂施工完成之后发挥作用,并且要快速破胶,缩短时间内返排。在聚合物乳液使用过程中,还可能出现其他相关问题,如残渣降解和离子灵敏性问题。

7.1 破乳剂基本内容

7.1.1 性能需求

原油乳化液的破乳剂应具备以下性能:
(1)将原油快速分解为水和残留水最少的油;
(2)保质期长;
(3)制备容易。

7.1.2 破乳机理

7.1.2.1 油水乳化剂的稳定性

油水乳化剂的稳定性主要由原油沥青胶体和树脂的界面层决定,只有破乳剂才能打破这种界面。在体系中加入水溶性破乳剂,最初界面上的乳化稳定剂会被破乳剂取代。此外,加入非活性化合物,润湿性也可能发生改变。相反,除了原油胶体置换外,油溶性破乳剂的机理是,添加的破乳剂形成的稳定效果与界面破裂引起破乳中和作用(Kotsaridou - Nagel 和 Kragert,1996)。

7.1.2.2 界面张力弛豫

原油破乳剂的有效性和剪切黏度的降低与油水界面的动态张力梯度相关。若体系中含有有效的破乳剂,界面张力(IFT)弛豫发生很快。弛豫时间短表明 IFT 梯度被抑制。示踪破乳剂的电子自旋共振实验表明,破乳剂在大量的油中形成了反胶束簇(Mukherjee 和 Kushnick,1987)。在张力弛豫过程中,界面上的破乳剂的缓慢释放是反应速率的决定步骤。

7.2 化学试剂

压裂液中的破乳剂用量通常在 0.1%~0.5%(体积分数)之间(Crews,2006)。常用破乳剂化学试剂种类见表 7.1。

表 7.1 破乳剂种类

种类	种类
烷基硫酸盐和磺酸盐	乙氧基醇
烷基膦酸酯	有机和无机铝盐
烷基季铵和氧化胺	丙烯酸酯—表面活性剂的共聚物
烷氧基化的聚亚烷基聚(胺)	丙烯酸酯—树脂共聚物
脂肪酸聚烷基芳基氯化铵	丙烯酸酯—烷基芳胺的共聚物
聚亚烷基乙二醇和醚	羧基化合物—多元醇
聚丙烯酸酯和丙烯酰胺	带有乙烯基化合物的烷氧基化物的共聚物或三元共聚物
烷基酚树脂	单胺缩合物或低聚胺烷氧基化物
低聚胺烷氧基化物	二羧酸和氧化烯嵌段共聚物
烷氧基羧酸酯	

另外,还有一些化学试剂能够提高破乳剂性能,称之为破乳剂增强剂,包括:醇、芳族化合物、链烷醇胺、羧酸、羧酸胺、亚硫酸氢盐、氢氧化物、硫酸盐、磷酸盐、多元醇及其混合物(Crews,2006)。

在油气增产工作液中加入烷基醚,有机酸酯或多元醇,对产油层中的泡沫和乳化液起到破乳和消泡作用(Leshchyshyn 等,2012;Giffin,2012)。这种作用取决于时间和温度,因此要现场控制(Crews,2007;Manz 等,2005)。含有破乳剂的特定混合物已有介绍(Crews,2007;Manz 等,2005)。

7.3 螯合剂

在水基压裂液中,可生物降解的且无毒的螯合剂成分有很多功能,如:
(1)破乳剂;
(2)破乳剂增强剂;

（3）阻垢剂；

（4）交联延迟剂；

（5）温度稳定剂；

（6）酶破胶剂稳定剂。

适宜的螯合剂见表7.2。

表7.2 螯合剂（Crews,2006）

化合物	化合物
聚天冬氨酸钠	糖醇
亚氨基二酸钠	单糖
羟亚乙基二钠亚氨基二乙酸	二糖
葡萄糖酸钠	

文献中的商品名称		
商品名称	描述	供应商
Aldacide® G	杀菌剂,戊二醛（Giffin,2012）	Halliburton Energy Services, Inc.
Envirogem®	非离子表面活性剂（Giffin,2012）	Air Products and Chemicals, Inc.
Rhodoclean™	非离子表面活性剂（Giffin,2012）	Rhodia Inc. Corp.

参 考 文 献

Crews, J. B., 2006. Biodegradable chelant compositions for fracturing fluid. US Patent 7 078 370. assigned to Baker Hughes Incorporated, Houston, TX, 18 July 2006. < http://www.freepatentsonline.com/7078370.html >.

Crews, J. B., 2007. Fracturing fluids for delayed flowback operations. US Patent 7 256 160. Assignedto Baker Hughes Incorporated, Houston, TX, 14 August 2007. < http://www.freepatentsonline.com/7256160.html >.

Crowell, R. F., 1984. Formation fracturing method. US Patent 4 442 897. assigned to Standard Oil Company, Chicago, IL, 17 April 1984. < http://www.freepatentsonline.com/4442897.html >.

Giffin, W. J., 2012. Compositions and processes for fracturing subterranean formations. US Patent 8 293 687. assigned to Titan Global Oil Services Inc., Bloomfield Hills, MI, 23 October 2012. < http://www.freepatentsonline.com/8293687.html >.

Graham, J. W., Gruesbeck, C., Salathiel, W. M., 1976. Method of fracturing subterranean formationsusing oil – in – water emulsions. US Patent 3 977 472. assigned to Exxon Production ResearchCompany, Houston, TX, 31 August 1976. < http://www.freepatentsonline.com/3977472.html >.

Kotsaridou – Nagel, M., Kragert, B., 1996. Demulsifying water – in – oil – emulsions through chemicaladdition (Spaltungsmechanismus von Wasser – in – Erdöl – Emulsionen bei Chemikalienzusatz). Erdöl Erdgas Kohle 112(2), 72 – 75.

Leshchyshyn, T. T., Beaton, P. W., Coolen, T. M., 2012. Hydrocarbon – based fracturing fluidcompositions, methods of preparation and methods of use. US Patent 8 211 834. Assignedto Calfrac Well Services Ltd. Alberta, CA, 3 July 2012. < http://www.freepatentsonline.com/8211834.html >.

Manz, D. H., Mahmood, T., Khanam, H. A., 2005. Oil and gas well fracturing (frac) water treatment process. US

Patent Application 20050098504. assigned to Davnor WaterTreatment Technologies Ltd. ,Calgary,CA,12May 2005. <http://www.freepatentsonline.com/20050098504.html>.

Mukherjee,S. ,Kushnick,A. P. ,1987. Effect of demulsifiers on interfacial properties governingcrude oil demulsification. In:Proceedings Volume. Annual Aiche Meeting,New York,15 – 20 November 1987.

Tambe,D. ,Paulis,J. ,Sharma,M. M. ,1995. Factors controlling the stability of colloid – stabilizedemulsions:Pt. 4: evaluating the effectiveness of demulsifiers. J. Colloid Interface Sci. 171(2),463 – 469.

8 黏土稳定剂

在石油运移过程中,页岩经常会带来一些问题。在20世纪50年代初期,许多土壤力学专家对黏土的膨胀感兴趣。在钻井过程中保持井筒稳定性是十分重要的,压裂作业过程中也是如此,尤其是在具有水敏性的页岩和黏土地层中。

这些类型的岩石地层会吸收压裂中的液体。从而致使岩石膨胀,可能导致井筒崩塌。文献中已经评估了黏土膨胀和由此可能出现的问题(Durand 等,1995a,1995b;Zhou 等,1995;Van Oort,1997;Conway 等,2011;Patel 和 Patel,2012)。表8.1给出了各种添加剂对黏土的稳定性能。

表8.1 黏土稳定剂的类型

添加剂类型	参考文献
聚合物晶格	Stowe 等(2002)
阴离子和阳离子单体的共聚物	
羟基醛或羟基酮	Westerkamp 等(1991)
多元醇及其碱性盐	Hale 和 van(1997)
季铵盐类化合物	
就地交联的环氧树脂	Coveney 等(1999a,b)
季铵盐类羧化物[BD,LT]	Himes(1992)
季铵化三羟基烷基胺[LT]	Patel 和 McLaurine(1993)
聚(乙烯醇)、硅酸钾和碳酸钾	Alford(1991)
苯乙烯与顺丁烯二酸酐(MA)取代物的共聚物	Smith 和 Balson(2000)
羧甲基纤维素钾	Palumbo 等(1989)
具有磺基琥珀酸酯衍生物表面活性剂,两性离子表面活性剂水溶性聚合物[BD,LT]	Alonso–Debolt 和 Jarrett(1995,1994)

注:BD—可生物降解,LT—低毒性,SF—增产措施工作液。

8.1 黏土的特征

黏土矿物本质上是晶体结构。黏土的晶体结构决定了其特征。典型的黏土具有片状云母型结构。黏土的层状结构是由晶体的晶片面对面堆叠而成的。每个晶片被称为单元层,单元层的表面被称为基底表面。一个单元层是由多个单体组成的。一种单体类型为八面体,由铝或镁原子八面体与羟基的氧原子配合形成。另一种单体是四面体,由硅原子与氧原子形成四面配位体。同一个单元层的单体通过共享氧原子连接在一起。

当一个八面体和一个四面体单体之间相连接时,一个基底表面含有暴露的氧原子,而另一个的基底表面有露出的羟基。两个四面体和一个八面体通过共享氧原子相连接也是很常见的。一个八面体夹在两个四面体之间,而形成的结构称为霍夫曼结构(Hoffmann 和 Lipscomb,1962)。因此,霍夫曼结构中的两个基底表面都是由露出的氧原子构成的。

单元层由较弱的引力面对面堆栈在一起。在相邻单元层对应位面之间的距离称为晶格间距(c – spacing)。由3个单体组成的单元层黏土晶体结构的晶格间距约 9.5×10^{-7} mm。

8.1.1 黏土膨胀

在黏土矿物晶体中,单体结构内原子通常有不同的化合价,致使在晶体表面显示出负电位。在这种情况下,表面可以吸附一个阳离子。这些被吸附的阳离子称为可交换阳离子,因此,当黏土晶体在水中时,这些阳离子可以与其他阳离子发生化学交换。此外,离子也可以吸附在黏土晶体边缘并与水中的其他离子发生交换(Patel 等,2007)。

在黏土晶体结构中和吸附在晶体表面上可交换阳离子的类型对黏土膨胀的影响很大。黏土膨胀是一种现象,水分子包围黏土晶体结构并且自身定位而增加结构的晶格间距,从而导致黏土矿物体积的增加。可能发生以下两种类型的膨胀(Patel 等,2007):

一种是表面水化,水分子吸附在晶体表面。一层水分子与暴露在晶格表面的氧原子以氢键相连。后续层的水分子在单元层之间形成准晶体结构,导致晶格增加。大多数类型的黏土膨胀都是这种方式。

另一种是渗透膨胀。黏土矿物单元层间的阳离子浓度高于周围水中的氧离子浓度,水渗透进单元层致使晶格间距增加。渗透膨胀导致的体积增加比表面水合作用更大。然而,只有少数黏土,如钠蒙皂石,以这种方式膨胀(Patel 等,2007)。

黏土是天然层状矿物,是火成岩风化和分解形成的。黏土矿物学的详细信息可以在相关专著中获得(Grim,1968;Murray,2007)。每一层都是由 Al^{3+},Mg^{2+} 或 Fe^{3+} 氧化物的正八面体和 Si^{4+} 氧化物的四面体混合而成(Auerbach,2007)。如果黏土矿物是由一个四面体和八面体构成的,称之为1:1黏土。如果黏土是两个四面体中间夹一个八面体,则称之为2:1黏土。

八面体和四面体层如图8.1所示。黏土晶格中的金属原子可以适当地被置换,导致个别黏土层整体带负电荷。

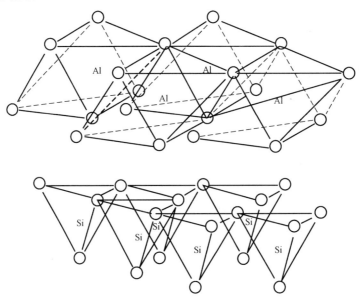

图8.1 黏土中的八面体和四面体晶层(Murray,2007)

这个电荷由位于层间区域的阳离子补偿。后者可以自由交换阳离子。黏土矿物阳离子的交换能力取决于晶体大小、pH 值和阳离子类型。这些阳离子不仅仅可以是小离子,也可以是聚阳离子(Blachier 等,2009)。

已经有关于黏土上聚季铵盐聚氧离子的吸附研究。在一定范围内,可以观察到聚季铵盐的吸附曲线和钠的释放曲线对应叠加,如图 8.2 所示。在低聚合物浓度时,替换的反离子胺聚合物几乎是 1∶1 的关系。此外,黏土矿物的硅酸盐四面体表面相对疏水。这个特性可以使聚合物中的中性有机化合物嵌入。

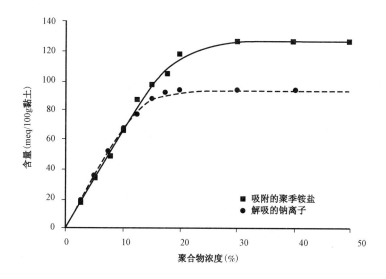

图 8.2　钠离子与聚季铵盐阳离子之间的交换(Blachier 等,2009)

蒙皂石是常见的典型的 2∶1 黏土(Anderson 等,2010)。饱和的钠蒙皂石的膨胀非常显著。这种膨胀造成在油井作业时页岩的敏感性。在最坏的情况下,井筒会因黏土膨胀而崩塌。

据报道,黏土矿物中可交换阳离子对膨胀量有很大影响。在黏土结构中的有效反应位置上,可交换的阳离子与水分子相竞争。一般来说,高价阳离子比相应的低价阳离子更容易吸附。因此,具有低价可交换阳离子的黏土比具有高价可交换阳离子的黏土更容易膨胀。

减少黏土膨胀是一个重要的研究方向。为了有效地降低黏土膨胀度,需要弄清楚膨胀的机理。基于这一认识,高效防膨剂得到了迅速发展。

合适的黏土防膨剂必须显著降低黏土的水化作用,同时必须满足日益严格的环境保护要求。

膨胀以离散的方式发生,即逐步形成完整的水合物。层间的距离过渡与相变的热力学过程类似。电渗膨胀只能发生在层间具有可交换阳离子的黏土矿物中。这种类型的膨胀要比晶体的膨胀大得多。

饱和的钠蒙皂石具有强烈的电渗膨胀倾向。相比之下,饱和的钾蒙皂石不会以这样的方式膨胀。因此,适当的离子交换反应可以有助于黏土稳定(Anderson 等,2010)。

夹有可交换碱金属阳离子的蒙皂石的水解吸等温线显示,较大的阳离子吸附的水更少(Mooney 等,1952)。此外,膨胀趋势与阳离子的水化能具有一定关系(Norrish,1954)。硬黏土因其具有较大的膨胀能力而闻名(Klein 和 Godinich,2006)。

8.1.2 蒙皂石

蒙皂石黏土,如膨润土和高岭石黏土,也适合作为水包油乳液的固体稳定剂。膨润土易于脱落(Bragg 和 Varadara,2006)。作为矿物,膨润土是由颗粒自然聚集组成的,可以分散在水中,可以剪切破碎成到平均粒径 $2\mu m$ 或更小的单颗粒。然而,每个颗粒是由含有大约100层的硅酸盐单元叠加而成的,这种硅酸盐单层之间由大约1nm厚的夹杂原子连接,如钙。

通过原子交换,如钙被钠或锂原子交换,这一厚度将更大,在淡水中具有更强的吸收水分子的能力,然后膨润土在淡水中,可以破碎分散成1nm厚的薄层,称之为基本粒子。这种分层过程的结果是形成由精细分散膨润土种组成的凝胶(Bragg 和 Varadaraj,2006)。

8.1.3 基本原则

在文献中提供了几个文件,可以作为实际设计中,选择适当的黏土稳定体系或完成井筒稳定性分析的基本原则(Chen 等,1996;Crowe(1990,1991a,1991b);Evans 和 Ali,1997;Scheuerman 和 Bergersen,1989)。

8.2 导致不稳定的机理

在各种井筒作业中,页岩稳定是一个重要问题。稳定问题经常是由于页岩膨胀造成的。已经有几种机制可以解释(Gazaniol 等,1994,1995)。这些机制包括:孔隙压力扩散、塑性、各向异性、毛细管效应、渗透性和物理化学变化。最重要的是,一般认为导致页岩不稳定的是以下三个过程(Bailey 等,1994):

(1)液体在井筒和页岩之间的运动(限于从井筒流入页岩)。

(2)在页岩与滤失液之间的相互作用期间发生应力和应变的变化。

(3)由于钻井液滤液的浸入,导致的软化和刻蚀以及随之发生的页岩内部的化学变化。

这些效应中的主要原因是化学性质,即黏土的水化作用。即使在大多数抑制液体中,如油基钻井液,也会观察到井眼不稳定现象。这表明机械方面的影响也很重要。事实上,一般认为是化学和机械共同作用的结果。基于此原因,在某些加载条件下,仍然难以预测中等到深层的岩石行为。

页岩的稳定性取决于页岩中的传质过程(如湍流、渗流、离子扩散、压力)和化学变化(如离子交换、含水变化、膨胀压力)之间的复杂关系。

黏土或页岩具有吸水的能力,造成井的不稳定或者是因为某些矿物的膨胀,或者是因为孔隙压力变化抵消了支撑压力。水基液体与页岩的反应取决于其初始的水活度和液体组成。

页岩的行为可以分为变形机制或传质机制(Tshibangu 等,1996)。钻井液的矿化度、密度和滤饼性质优化是一个重要问题。

8.2.1 黏土膨胀动力学

已经有研究人员研究了膨胀动力学(Suratman,1985)。研究了纯黏土(蒙皂石、伊利石和高岭石)和聚合抑制剂,建立了表观动力学规律。

8.2.2 水化应力

由化学势引起的应力,如水化应力,对井眼稳定性的影响相当大(Chen 等,1995)。当总压力和水的化学势增加,水被吸收到黏土晶格间。

若可以自由移动,就会造成在晶格间移动距离增加(膨胀),若膨胀受限,就会产生水化应力(Tan 等,1997)。水化应力增加,导致了孔隙压力的增加以及随后的钻井液有效支撑降低,从而导致了井筒不稳定。

8.2.3 井筒稳定模型

已经开发出的井筒稳定模型,考虑了钻井液与页岩之间的机械和化学相互作用(Mody 和 Hale,1993)。该模型综合考虑了化学诱导应力与机械诱导应力,基于钻井液与页岩摩尔自由能在水中的热力学差异的化学诱导应力变化,与机械诱导应力共同作用。基于这个模型,可以获得最佳的钻井液密度和盐浓度。

更新的是基于表面积的稳定模型,平衡水含量与压力的关系,以及双电层理论可以成功地表征井壁稳定性问题(Wilcox,1990)。表面积、膨胀压力和固体需水量模型的应用,可以综合考虑膨胀模型与钻井液控制过程方法,从而改进水基钻井液在活跃的或陈旧页岩层中的设计。

8.2.4 水基钻井液的页岩抑制

聚合物可以稳定页岩的一个潜在机理是降低了水侵入页岩的速度。水侵的控制不是与页岩稳定相关的唯一机理(Ballard 等,1993),还有聚合物添加剂的影响。渗透现象影响了水在页岩中的流动速度。

8.2.5 抑制活性泥质地层

泥质地层非常容易与水反应。这样的地层可以用具有亲水和疏水链接的聚合物溶液进行稳定(Audibert 等,1993a,1993b)。亲水部分由聚氧化乙烯组成,疏水端基来自异氰酸酯。由于其具有吸附和疏水的能力,该聚合物可以抑制泥质岩的膨胀或分散。

8.2.6 液体对地层的伤害

由于钻井液侵入地层造成的伤害是钻井中众所周知的问题。钻井液侵入地层一般是由于静水柱压力大于地层压力,形成压差造成的,特别是在低压力或衰竭地层(Whitfill 等,2005)。

钻开岩石和液体穿透过岩石的能力也会导致钻井液侵入地层。当在过平衡条件下,钻开枯竭砂层时,钻井液会逐渐渗透进入地层,除非在井筒壁上存在有效的流动障碍。

8.3 膨胀抑制剂

膨胀抑制剂的作用方式是化学的,而不是机械方式。它们改变了离子强度和液体在黏土中的传递行为。阳离子和阴离子两者对于黏土膨胀的抑制都是很重要的(Doleschall 等,1987)。

8.3.1 盐

可以添加 KCl 抑制膨胀,需要相对较高的浓度。不带电荷的聚合物和聚(电解质)是另两种膨胀抑制剂(Anderson 等 2010)。

8.3.2 季铵盐

胆碱盐类是欠平衡钻井作业中有效的防膨剂(Kippie 和 Gatlin,2009)。

胆碱盐类称之为季铵盐,含有 N,N,N-三甲基乙醇胺阳离子。胆碱卤化物离子盐的一个例子就是氯化胆碱。

制备 8.1:三乙醇胺甲基氯的制备方法是在三乙醇胺水溶液中加入过量的氯甲烷并加热数小时。在反应完成后,将过量的氯甲烷蒸发掉。

胆碱甲酸盐是由甲酸与胆碱水溶液在搅拌条件下制备的。

黏土地层含有黏土颗粒。如果在这种地层中使用水基钻井液,就会发生离子交换和水化作用等。这些反应会导致黏土颗粒的膨胀、破碎或分散。最后,井壁会脱落甚至完全崩塌(Eoff 等,2006)。

某些添加剂可以阻止这种反应出现。这些添加剂是季铵化聚合物。已在实验室试验中证明这种聚合物可以大大减少侵蚀页岩。季铵化聚合物可以按如下步骤合成得到(Eoff 等,2006):

(1)卤代烷与 AA 基胺衍生物进行季铵化反应,随后进行聚合反应;或

(2)先聚合反应,然后聚合物基团进行季铵化反应。

制备 8.2:季铵盐单体可以用二甲氨基甲基丙烯酸乙酯与十六烷基溴化铵混合制备。将混合物加热到 43℃ 并搅拌 24h 后,将混合物倒入石油醚中,其中季铵化单体沉淀析出(Eoff 等,2006)。该反应如图 8.3 所示。

共聚物可以用上述季铵盐单体和二甲氨基甲基丙烯酸乙酯制备。水溶液用硫酸完全中和,生成 $2,2'$-偶氮聚(2-脒基丙烷)二盐酸盐,参见图 8.4。这种引发剂是水溶性的。聚合作用在 43℃ 恒温条件下经过 18h 后完成(Eoff 等,2006)。

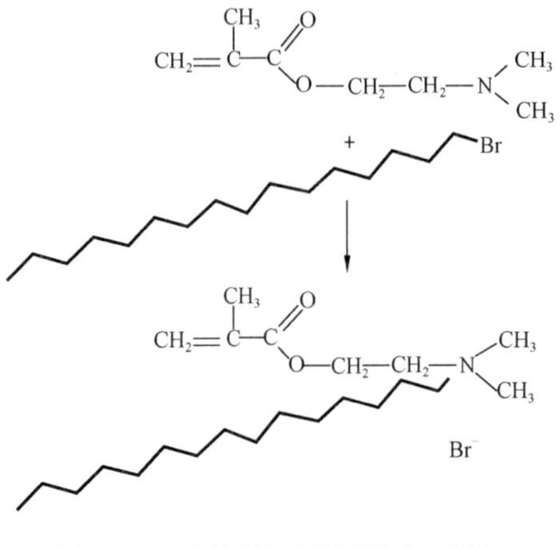

图 8.3 二甲氨基甲基丙烯酸乙酯与十六烷基溴化铵的季铵化反应

图 8.4 水溶性自由基引发剂

由二甲氨基甲基丙烯酸乙酯得到的季铵化聚合物也有描述。在二甲氨基甲基丙烯酸乙酯合成的均聚物的水溶液中,加入盐酸钠调节 pH 值至 8.9。然后再添加一些水,十六烷基溴化铵作为烷基化剂,苯甲基十六烷基二甲基溴化铵为乳化剂。然后在 60℃ 搅拌并加热该混合物 24h(Eoff 等,2006)。

8.3.3 甲酸钾

在钻井和作业中可以使用甲酸钾稳定黏土。此外,添加阳离子地层控制添加剂。甲酸钾可以用氢氧化钾和甲酸就地生成。阳离子添加剂基本上是含季铵盐的聚合物,例如,二烯丙基二甲基氯化铵或丙烯酰胺聚合物(AM)(Smith,2009)。

在黏土压块流量实验中,在一个给定的时间内,流过的量越高表明黏土稳定性越好,加入少量的甲酸钾,可以增加给定浓度聚合物的通过量。例如,在配方中加入0.1%的聚二烯丙基二甲基氯化铵,10min 流出的体积达到112mL。

相同的聚合物,当加入甲酸钾后,聚合物浓度仅需原来聚合物浓度的一半,即0.05%,流出体积就可以达146mL,说明甲酸钾具有较好的黏土稳定性能,另外一个可能是在加入甲酸钾后的协同作用(Smith,2009)。

8.3.4 糖类衍生物

甲基葡萄糖苷和环氧烷,如环氧乙烷(EO)、环氧丙烷(PO),或1,2 - 环氧丁烷的反应产物可以作为黏土稳定剂。这种添加剂在室温条件下易溶于水,但在高温下不溶于水(Clapper 和 Watson,1996)。由于其在高温下不溶于水,这些化合物会浓聚在钻头切削表面,钻孔表面和钻屑表面等重要表面上。

8.3.5 磺化沥青

沥青是一种固体,随着组分不同颜色从黑棕色到黑色。在加热时变软,冷却时重新变硬。沥青是不溶于水的,在水中很难分散或乳化。

可以通过沥青与硫酸和三氧化硫反应得到磺化沥青。用碱金属类氢氧化物如氢氧化钠或NH_3中和,得到磺酸盐。用热水仅可以提取有限量的磺化产品。不过,可以得到的这部分是水溶性的,其质量是至关重要的。

磺化沥青通常用于水基和油基钻井液(Huber 等,2009)。磺化沥青作为黏土稳定剂的作用机理是:带负电的磺化高分子吸附在黏土晶格的阳离子端,形成中性屏障,可以抑制黏土吸收水分。

此外,由于磺化沥青是部分亲油的,因此疏水,纯粹的物理原因制约了水侵入黏土。正如前文已经提到的,在水中的溶解度是磺化沥青正确应用的关键。由于水溶性阴离子聚合物组分的引入,水不溶性沥青的比例可以显著降低。

换句话说,通过引入聚合物成分,水溶性部分的比例增加。特别适合的是木质素磺酸盐、磺化苯酚、萘、酮、丙酮和氨基增塑树脂等(Huber 等,2009)。

8.3.6 接枝共聚物

研究了苯乙烯和MA 接枝聚(乙二醇)(PEG)共聚物的黏土稳定性能(Smith 和 Balson,2004)。

用滚动实验中页岩回收率确定对页岩的抑制能力。试验使用 Oxford 黏土岩屑,是水敏性页岩,颗粒范围2~4mm。膨胀试验在7.6% KCl 水溶液中进行。

接枝共聚物用的是苯乙烯和MA 共聚物。用不同相对分子质量的聚乙二醇(PEG)与聚合物接枝。不同类型的PEG 的页岩回收率见表8.2。

表8.2 页岩回收率(Smith 和 Balson,2004)

样品	KCl(%)	页岩回收率(%)
仅 KCl	7.6	25
PEG	7.6	38
SMAC MPEG 200	7.6	54
SMAC MPEG 300	7.6	87
SMAC MPEG 400	7.6	85
SMAC MPEG 500	7.6	72
SMAC MPEG 600	7.6	69
SMAC MPEG 750	7.6	70
SMAC MPEG 1100	7.6	66
SMAC MPEG 1500	7.6	49
仅 KCl	12.9	27
PEG	12.9	53
SMAC MPEG 500	12.9	85
SMAC2∶1MPEG 500	12.9	95

注:SMAC 为苯乙烯与 MA 共聚物;
SMAC 2∶1为苯乙烯与 MA 共聚物,每个 MA 对应2个苯乙烯;
MPEG 为聚乙二醇甲醚,数字为参考相对分子质量。

对于接枝 PEG 相对分子质量,似乎有一个最佳值。此外,表8.2中下部数据表明,增加骨架中苯乙烯的量也会增加页岩的回收率。

8.3.7 聚氧化烯胺

降低黏土膨胀的一种方法是加入盐溶液。盐一般可以降低黏土膨胀。然而,盐会絮凝黏土,致使流体大量滤失并且几乎完全丧失触变性。此外,增加含盐量通常会降低其他钻井液添加剂的性能(Patel 等,2007)。

用于控制黏土膨胀的另一种方法是,在钻井液中使用有机页岩抑制剂。一般认为有机页岩抑制剂的分子可以吸附在黏土表面上,占领黏土表面水分子的活性位点,从而起到降低黏土膨胀的作用。

聚氧化烯胺是伯氨基团连接到聚醚骨架上的一大类化合物。还称为聚醚胺,有多种不同的分子质量,最高可达5kDa。

聚氧化烯二胺也可以作为页岩抑制剂。可以在氨基化合物存在条件下,由环氧乙烷化合物开环聚得到。该化合物可以由聚醚胺与两当量的环氧乙烷反应合成。或者,由 PO 与二胺反应得到(Patel 等,2007)。聚醚骨架是 EO,PO 或者这些环氧化合物的混合物(Patel 等,2007)。

一个典型的聚醚胺如图8.5所示。此类产品属于 Jeffamine® 系列产品。相应的页岩水化抑制剂是 N-烷基化的 2,2′-二氨基乙醚。

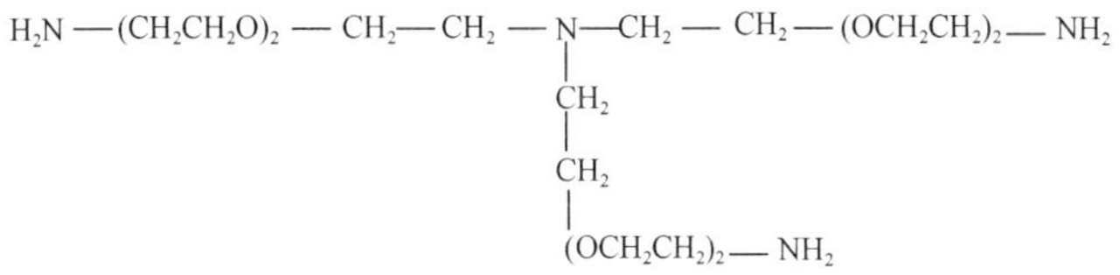

图 8.5　聚醚胺(Klein 和 Godinich,2006)

8.3.8　阴离子聚合物

阴离子聚合物是有效的黏土稳定剂,通过长链上的负离子吸附到黏土颗粒的正电位上,或者通过氢键吸附在水化的黏土粒子表面上达到稳定黏土的目的(Halliday 和 Thielen,1987)。当聚合物附着在黏土表面上,表面水化作用会降低。

表面保护层密封或限制了表面裂缝和孔隙,从而减少或预防由于虹吸现象进入页岩的滤液。这个稳定过程通过 PAC 得到加强。氯化钾提高了聚合物在黏土上的吸附速度。

8.3.9　顺丁烯二酰亚胺的铵盐

含 MA 聚合物亚胺的铵盐混合物常用作黏土稳定。例如,在乙二醇(EG)溶液中,用 MA 与二胺,如二甲氨基丙胺,反应制备得到这些类型的盐(Poelker 等,2009)。主要的氮二甲酰亚胺形成亚胺键。

此外,还可以添加到 MA 的双键上。而且 EG 也可以加到双键上,但其自身也可能会浓缩成酸酐。重复这些反应,可以形成低聚物的混合物。基本的反应如图 8.6 所示。

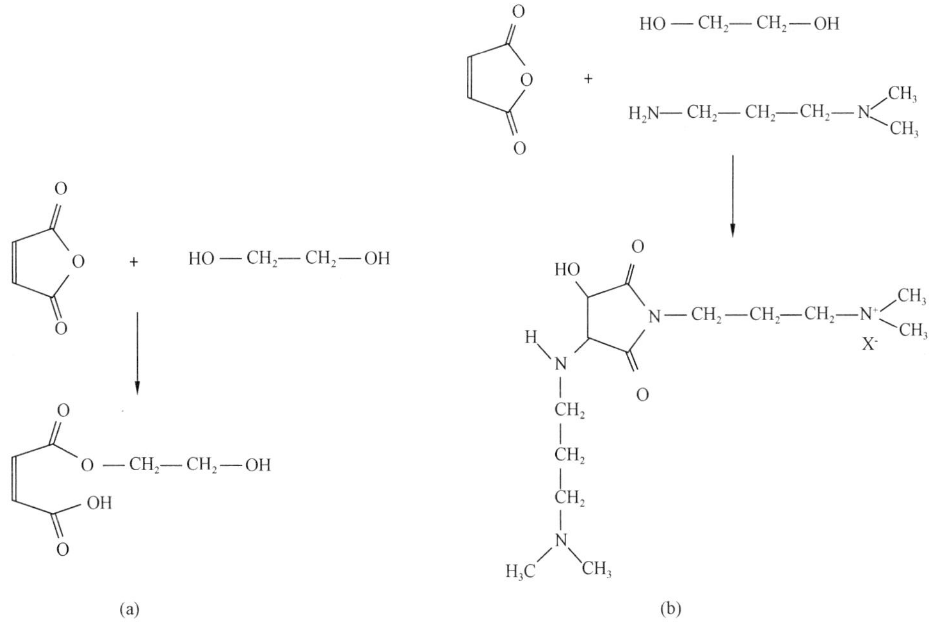

图 8.6　与乙二醇缩合开始(a)和亚胺铵盐的形成(b)(Poelker 等,2009)

最后,用乙酸或甲酸中和该产品至 pH 值为 4。用 Bandera 砂岩测试其性能。用甲磺酸进行中和的比用乙酸中和的稍微少一点。这些材料特别适用于水基压裂液。

8.3.10 胍基共聚物

可以添加胍基共聚物降低黏土膨胀和颗粒运移(Murphy 等,2012)。该共聚物是一个氨基、甲醛、一个亚烷基胺和无机或有机酸铵盐的缩合产物。制备的基本过程已有报道(Waldmann,1997),制备方法摘要见制备 8.3。

制备 8.3:在搅拌条件下,将二乙烯三胺加热到 55~60℃,加入氯化铵。然后将混合物加热到 95~100℃,添加二甘醇和双氰胺。逐渐升高反应器温度至 195℃,用特殊的方案冷却(Waldmann,1997)。

单体的变化见表 8.3。胍基共聚物具有较好分子质量 1~20kDa。例如,胍基共聚物可减少添加到压裂液中降阻剂总量的 30%~70%(Murphy 等,2012)。比传统的化合物显著的节约成本。

表 8.3 胍基共聚物单体(Murphy 等,2012)

反应的胺类	反应的羰基类	反应的胺类
二氰二胺	甲醛	氨乙基哌嗪
胍胺	多聚甲醛	乙二胺
胍	尿素	二乙烯三胺
三聚氰胺	硫脲	三乙烯四胺
磷酰氰胺	乙二醛	丙二胺
脒基脲	乙醛	三丁基烯甲基四胺
脒基硫脲	丙醛	四丁基烯五胺
烷基胍	丁醛	烯二戊酯三胺
芳基胍	戊二醛丙酮	三戊烯甲基四胺戊乙二胺

8.3.11 特殊的黏土稳定剂

在黏土为主导的地层中进行的作业量增加,促进了众多的黏土稳定剂处理和添加剂的发展。使用的大多数添加剂是有机阳离子高分子聚合物。然而,已有研究证明,这些稳定剂在低渗透地层不太有效(Himes 等,1989)。

在石油钻井、完井和修井过程中,氯化钾和氯化钠等盐类,作为暂时的黏土稳定剂,已经使用多年。由于盐的使用量大,且稳定区域小以及对环境的潜在危害,许多服务商正在寻找替代品。

最近的研究揭示了各种离子的物理性质(例如,K^+,Na^+)与其作为暂时黏土稳定剂的效率的关系。这些属性用于指导合成有机阳离子(表 8.4),生产高效的黏土稳定剂。

表 8.4 特殊的黏土稳定剂

化合物	参考文献
氯化铵	
氯化钾①	Yeager 和 Bailey(1988)
二甲基己二烯铵盐②	Thomas 和 Smith(1993)
N-烷基吡啶卤化物	
N,N,N-三烷基铵苯的卤化物	
N,N-二烷基吗啉的卤化物③	Himes(1992)和 Himes 和 Vinson(1989)
MA 与烷基二胺的均聚物反应产品④	Schield 等(1991)
四甲基氯化铵和氯甲烷	Aften 和 Gabel(1992)
乙烯胺的季铵盐缩聚物④	
季铵类化合物⑤	Hall 和 Szememyei(1992)

① 添加到用柴油基的凝胶中。
② 黏土防膨最小浓度 0.05%。
③ 烷基为甲基、乙基、丙基、丁基烷。
④ 协同作用阻碍地层黏土吸水。
⑤ 羟基取代烷基的半径。

文献中的商品名称

商品名称	描述	提供者
Aerosil®	气相二氧化硅(Bragg 和 Varadaraj,2006)	Degussa AG
Carbolite™	不同粒径的陶粒支撑剂(Kippie 和 Gatlin,2009)	Carbo Corp.
Dacron®	聚对苯二甲酸乙二醇酯(Kippie 和 Gatlin,2009)	DuPont
Jeffamine® D-230	聚氧化丙烯二胺(Klein 和 Godinich,2006)	Huntsman
Jeffamine® EDR-148	三甘醇二胺(Klein 和 Godinich,2006)	Huntsman
Jeffamine® HK-511	聚氧化烯胺(Klein 和 Godinich,2006)	Huntsman
Shale Guard™ NCL100	页岩防膨剂(Kippie 和 Gatlin,2009)	Weatherford Int.

由于可以在较低的盐浓度条件下使用,这些添加剂在酸化和压裂过程中,具有更多的优势(Himes 等,1990;Himes 和 Vinson,1991)。液体产品更容易处理和运输,与环境是相容的并且可以生物降解。

参 考 文 献

Aften,C. W.,Gabel,R. K.,1992. Clay stabilizing method for oil and gas well treatment. US Patent5 099 923,assigned to Nalco Chemical Co.,31 March 1992.

Alford,S. E.,1991. North Sea field application of an environmentally responsible water – base shalestabilizing system. In:Proceedings Volume. SPE/IADC Drilling Conference,Amsterdam,the Netherland,11 – 14 March 1991,pp. 341 – 355.

Alonso – Debolt,M. A.,Jarrett,M. A.,1994. New polymer/surfactant systems for stabilizingtroublesome gumbo shale. In:Proceedings Volume. SPE International Petroleum Conferenceof Mexico,Veracruz,Mexico,10 – 13 October 1994,pp. 699 – 708.

Alonso-Debolt, M., Jarrett, M., 1995. Synergistic effects of sulfosuccinate/polymer system forclay stabilization. In: Proceedings Volume. Vol. PD-65. Asme Energy-Sources Technological Conference Drilling Technological Symposium, Houston, 29 January -1 February 1995, pp. 311-315.

Anderson, R. L., Ratcliffe, I., Greenwell, H. C., Williams, P. A., Cliffe, S., Coveney, P. V., 2010. Clay swelling—A challenge in the oilfield. Earth-Sci. Rev. 98(3-4), 201-216. < http://www.sciencedirect.com/science/article/B6V62-4XRCRVK-1/2/2ff8baefa17f8a009368b42d7f32f3ad >.

Audibert, A., Lecourtier, J., Bailey, L., Maitland, G., 1993a. Method for inhibiting reactiveargillaceous formations and use thereof in a drilling fluid. WO Patent 9 315 164, assignedto Schlumberger Technol. Corp., Schlumberger Serv. Petrol, and Inst. Francais Du Petrole, 5 August 1993.

Audibert, A., Lecourtier, J., Bailey, L., Maitland, G., 1993b. Process for inhibiting reactiveargillaceous formations and application to a drilling fluid (procede d'inhibition de formationsargileuses reactives et application a un fluide de forage). FR Patent 2 686 892, assigned to Inst. Francais Du Petrole and Schlumberger Cambridge Re, 6 August 1993.

Audibert, A., Lecourtier, J., Bailey, L. C., Maitland, G., 1994. Use of polymers having hydrophilicand hydrophobic segments for inhibiting the swelling of reactive argillaceous formations (l'utilisation d'un polymere en solution aqueuse pour l'inhibition de gonflement des formationsargileuses reactives). EP Patent 578 806, assigned to Schlumberger Serv. Petrol and Inst. Francais Du Petrole, 19 January 1994.

Audibert, A., Lecourtier, J., Bailey, L., Maitland, G., 1997. Method for inhibiting reactiveargillaceous formations and use thereof in a drilling fluid. US Patent 5 677 266, assigned toInst. Francais Du Petrole, 14 October 1997.

Auerbach, S. M. (Ed.), 2007. Handbook of layered materials, reprint from 2004 edition. CRC Press, Boca Raton.

Bailey, L., Reid, P. I., Sherwood, J. D., 1994. Mechanisms and solutions for chemical inhibition of shale swelling and failure. In: Proceedings Volume. Recent Advances in Oilfield Chemistry, 5th Royal Society of Chemistry. International Symposium, Ambleside, England, 13-15 April 1994, pp. 13-27.

Ballard, T., Beare, S., Lawless, T., 1993. Mechanisms of shale inhibition with water based muds. In: Proceedings Volume. IBC Technical Services Ltd., Preventing Oil Discharge from Drilling Operation—The Options Conference, Aberdeen, Scotland, 23-24 June 1993.

Blachier, C., Michot, L., Bihannic, I., Barrès, O., Jacquet, A., Mosquet, M., 2009. Adsorption ofpolyamine on clay minerals. J. Colloid Interface Sci. 336(2), 599-606. < http://www.sciencedirect.com/science/article/B6WHR-4W2NDRN-B/2/0cdcd6ee0039c2579d4807db98e548d5 >.

Bragg, J. R., Varadaraj, R., 2006. Solids-stabilized oil-in-water emulsion and a method for preparingsame. US Patent 7 121 339, assigned to ExxonMobil Upstream Research Company, Houston, TX, 17 October 2006. < http://www.freepatentsonline.com/7121339.html >.

Chen, M., Chen, Z., Huang, R., 1995. Hydration stress on wellbore stability. In: ProceedingsVolume. 35th US Rock Mech Symposium, Reno, NV, 5-7 June 1995, pp. 885-888.

Chen, X., Tan, C. P., Haberfield, C. M., 1996. Wellbore stability analysis guidelines for practical welldesign. In: Proceedings Volume. SPE Asia Pacific Oil & Gas Conference, Adelaide, Australia, 28-31 October 1996, pp. 117-126.

Clapper, D. K., Watson, S. K., 1996. Shale stabilising drilling fluid employing saccharide derivatives. EP Patent 702 073, assigned to Baker Hughes Inc., 20 March 1996.

Conway, M., Venditto, J. J., Reilly, P., Smith, K., 2011. An examination of clay stabilization andflow stability in various north american gas shales. In: Proceedings of SPE Annual Technical Conference and Exhibition. SPE Annual Technical Conference and Exhibition, 30 October-2 November 2011, Denver, Colorado, USA, Society of Petroleum Engineers, Dallas, Texas, pp. 1-17. http://dx.doi.org/10.2118/147266-MS.

Coveney, P. V., Watkinson, M., Whiting, A., Boek, E. S., 1999a. Stabilising clayey formations. GBPatent 2 332 221, assigned to Sofitech NV, 16 June 1999.

Coveney, P. V. , Watkinson, M. , Whiting, A. , Boek, E. S. , 1999b. Stabilizing clayey formations. WO Patent 9 931 353, assigned to Sofitech NV, Dowell Schlumberger SA, and SchlumbergerCanada Ltd. , 24 June 1999.

Crowe, C. W. , 1990. Laboratory study provides guidelines for selecting clay stabilizers. In: Proceedings Volume. Vol. 1. Cim. Petroleum Soc/SPE International Technical Management, Calgary, Canada, 10 – 13 June 1990.

Crowe, C. W. , 1991a. Laboratory study provides guidelines for selecting clay stabilizers. SPEUnsolicited Pap SPE – 21556, Dowell Schlumberger.

Crowe, C. W. , 1991b. Laboratory study provides guidelines for selecting clay stabilizers. In: Proceedings Volume. SPE Oilfield Chemical International Symposium, Anaheim, California, 20 – 22 February 1991, pp. 499 – 504.

Doleschall, S. , Milley, G. , Paal, T. , 1987. Control of clays in fluid reservoirs. In: Proceedings Volume. 4th BASF AG et al Enhanced Oil Recovery Europe Symp. (Hamburg, Ger, 27 – 28October 1987), pp. 803 – 812.

Durand, C. , Onaisi, A. , Audibert, A. , Forsans, T. , Ruffet, C. , 1995a. Influence of clays onborehole stability: A literature survey: Pt. 1: Occurrence of drilling problems physico – chemicaldescription of clays and of their interaction with fluids. Rev. Inst. Franc. Pet. 50(2), 187 – 218.

Durand, C. , Onaisi, A. , Audibert, A. , Forsans, T. , Ruffet, C. , 1995b. Influence of clays on boreholestability: A literature survey: Pt. 2: Mechanical description and modelling of clays and shalesdrilling practices versus laboratory simulations. Rev. Inst. Franc. Pet. 50(3), 353 – 369.

Eoff, L. S. , Reddy, B. R. , Wilson, J. M. , 2006. Compositions for and methods of stabilizingsubterranean formations containing clays. US Patent 7 091 159, assigned to HalliburtonEnergy Services, Inc. , Duncan, OK, 15 August 2006. < http://www. freepatentsonline. com/7091159. html >.

Evans, B. , Ali, S. , 1997. Selecting brines and clay stabilizers to prevent formation damage. World Oil 218(5), 65 – 68.

Gazaniol, D. , Forsans, T. , Boisson, M. J. F. , Piau, J. M. , 1994. Wellbore failure mechanisms in shales: Prediction and prevention. In: ProceedingsVolume. Vol. 1. SPE Europe Petroleum Conference, London, UK, 25 – 27 October 1994, pp. 459 – 471.

Gazaniol, D. , Forsans, T. , Boisson, M. J. F. , Piau, J. M. , 1995. Wellbore failure mechanisms in shales: Prediction and prevention. J. Pet. Technol. 47(7), 589 – 595.

Grim, R. E. , 1968. Clay Mineralogy, second ed. McGraw – Hill, New York. Hale, A. H. , van Oort, E. , 1997. Efficiency of ethoxylated/propoxylated polyols with other additivesto remove water from shale. US Patent 5 602 082, 11 February 1997.

Hall, B. E. , Szememyei, C. A. , 1992. Fluid additive and method for treatment of subterraneanformations. US Patent 5 089 151, assigned to Western Co. North America, 18 February 1992.

Halliday, W. S. , Thielen, V. M. , 1987. Drilling mud additive. US Patent 4 664 818, assigned toNewpark Drilling Fluid In, 12 May 1987.

Himes, R. E. , 1992. Method for clay stabilization with quaternary amines. US Patent 5 097 904, assigned to Halliburton Co. , 24 March 1992.

Himes, R. E. , Vinson, E. F. , 1989. Fluid additive and method for treatment of subterraneanformations. US Patent 4 842 073, 27 June 1989.

Himes, R. E. , Vinson, E. F. , 1991. Environmentally safe salt replacement for fracturing fluids. In: Proceedings Volume. SPE East Reg Conference, Lexington, KY, 23 – 25 October 1991, pp. 237 – 248.

Himes, R. E. , Vinson, E. F. , Simon, D. E. , 1989. Clay stabilization in low – permeability formations. In: ProceedingsVolume. SPE Production Operation Symposium, Oklahoma City, 12 – 14 March 1989, pp. 507 – 516.

Himes, R. E. , Parker, M. A. , Schmelzl, E. G. , 1990. Environmentally safe temporary clay stabilizer foruse inwell service fluids. In: ProceedingsVolume. Vol. 3. Cim Petroleum Soc/SPE InternationalTechnical Management, Calgary, Canada, 10 – 13 June 1990.

Hoffmann, R. , Lipscomb, W. N. , 1962. Theory of polyhedral molecules. I. Physical factorizationsof the secular equation. J. Chem. Phys. 36(8),2179 – 2189.

Huber,J. ,Plank,J. ,Heidlas,J. ,Keilhofer,G. ,Lange,P. ,2009. Additive for drilling fluids. USPatent 7 576 039, assigned to BASF Construction Polymers GmbH,Trostberg,DE,18 August2009. < http://www. freepatentsonline. com/7576039. html >.

Kippie,D. P. ,Gatlin,L. W. ,2009. Shale inhibition additive for oil/gas down hole fluids and methodsfor making and using same. US Patent 7 566 686,assigned to Clearwater International,LLC,Houston,TX,28 July 2009. < http://www. freepatentsonline. com/7566686. html >.

Klein,H. P. ,Godinich,C. E. ,2006. Drilling fluids. US Patent 7 012 043,assigned to Huntsman Petrochemical Corporation,The Woodlands,TX,March 14 2006. < http://www. freepatentsonline. com/7012043. html >.

Mody,F. K. ,Hale,A. H. ,1993. A borehole stability model to couple the mechanics and chemistryof drilling fluid shale interaction. In:Proceedings Volume. SPE/IADC Drilling Conference,Amsterdam,Netherland 23 – 25 February 1993,pp. 473 – 490.

Mooney,R. W. ,Keenan,A. G. ,Wood,L. A. ,1952. Adsorption of water vapor by montmorillonite. II. Effect of exchangeable ions and lattice swelling as measured by X – ray diffraction. J. Am. Chem. Soc. 74(6),1371 – 1374.

Murphy,C. B. ,Fabri,J. O. ,Reilly Jr. ,P. B. ,2012. Treatment of subterranean formations. USPatent 8 157 010,assigned to Polymer Ventures,Inc. ,Charleston,SC,17 April 2012. < http://www. freepatentsonline. com/8157010. html >.

Murray,H. H. ,2007. Applied Clay Mineralogy:Occurrences,Processing,and Application of Kaolins,Bentonites,Palygorskite – Sepiolite,and Common Clays,vol 2. Elsevier,Amsterdam.

Norrish,K. ,1954. The swelling of montmorillonite. Discuss. Faraday Soc. 18,120 – 134.

Palumbo,S. ,Giacca,D. ,Ferrari,M. ,Pirovano,P. ,1989. The development of potassium cellulosicpolymers and their contribution to the inhibition of hydratable clays. In:Proceedings Volume. SPE Oilfield Chemical International Symposium,Houston,8 – 10 February 1989,pp. 173 – 182.

Patel,A. D. ,McLaurine,H. C. ,1993. Drilling fluid additive and method for inhibiting hydration. CA Patent 2 088 344,assigned to M I Drilling Fluids Co. ,11 October 1993.

Patel,M. A. ,Patel,H. S. ,2012. A review on effects of stabilizing agents for stabilization of weaksoil. Civil and Environmental Research 2(6),1 – 7.

Patel,A. D. ,Stamatakis,E. ,Davis,E. ,Friedheim,J. ,2007. High performance water based drillingfluids and method of use. US Patent 7 250 390,assigned to M – I L. L. C. ,Houston,TX,31 July2007. < http://www. freepatentsonline. com/7250390. html >.

Poelker,D. J. ,McMahon,J. ,Schield,J. A. ,2009. Polyamine salts as clay stabilizing agents. US Patent 7 601 675,assigned to Baker Hughes Incorp. ,Houston,TX,13 October 2009. < http://www. freepatentsonline. com/7601675. html >.

Scheuerman,R. F. ,Bergersen,B. M. ,1989. Injection water salinity,formation pretreatment,andwell operations fluid selection guidelines. In:Proceedings Volume. SPE Oilfield ChemicalInternational Symposium,Houston,8 – 10 February 1989,pp. 33 – 49.

Schield,J. A. ,Naiman,M. I. ,Scherubel,G. A. ,1991. Polyimide quaternary salts as clay stabilizationagents. GB Patent 2 244 270,assigned to Petrolite Corp. ,27 November 1991.

Smith,K. W. ,2009. Well drilling fluids. US Patent 7 576 038,assigned to Clearwater International,L. L. C. ,Houston,TX,August 18 2009. < http://www. freepatentsonline. com/7576038. html >.

Smith,C. K. ,Balson,T. G. ,2000. Shale – stabilizing additives. GB Patent 2 340 521,assigned toSofitech NV and Dow Chemical Co. ,23 February 2000.

Smith,C. K. ,Balson,T. G. ,2004. Shale – stabilizing additives. US Patent 6 706 667,16 March 2004. < http://www. freepatentsonline. com/6706667. html >.

Stowe, C., Bland, R. G., Clapper, D., Xiang, T., Benaissa, S., 2002. Water - based drilling fluids usinglatex additives. GB Patent 2 363 622, assigned to Baker Hughes Inc., 2 January 2002.

Suratman, I., 1985. A study of the laws of variation(kinetics) and the stabilization of swelling ofclay(contribution a l'etude de la cinetique et de la stabilisation du gonflement des argiles). Ph. D. Thesis, Malaysia.

Tan, C. P., Richards, B. G., Rahman, S. S., Andika, R., 1997. Effects of swelling and hydrationalstress in shales on wellbore stability. In: Proceedings Volume. SPE Asia Pacific Oil & GasConference, Kuala Lumpur, Malaysia, 14 - 16 April 1997, pp. 345 - 349.

Thomas, T. R., Smith, K. W., 1993. Method of maintaining subterranean formation permeability andinhibiting clay swelling. US Patent 5 211 239, assigned to Clearwater Inc., 18 May 1993.

Tshibangu, J. P., Sarda, J. P., Audibert - Hayet, A., 1996. A study of the mechanical andphysicochemical interactions between the clay materials and the drilling fluids: Applicationto the boom clay (Belgium) (etude des interactions mecaniques et physicochimiques entre lesargiles et les fluides de forage: Application a l' argile de boom (Belgique)). Rev. Inst. Franc. Pet. 51(4), 497 - 526.

Van Oort, E., 1997. Physico - chemical stabilization of shales. In: Proceedings Volume. SPE OilfieldChemical International Symposium, Houston, 18 - 21 February 1997, pp. 523 - 538.

Waldmann, J. J., 1997. Agents having high nitrogen content and high cationic charge based ondicyanimide dicyandiamide or guanidine and inorganic ammonium salts. US Patent 5 659 011, 19 August 1997. < http://www.freepatentsonline.com/5659011.html >.

Westerkamp, A., Wegner, C., Mueller, H. P., 1991. Borehole treatment fluids with clay swelling - inhibiting properties(ii) (bohrloch - behandlungsfluessigkeiten mit tonquellungsinhibierendeneigenschaften(ii)). EP Patent 451 586, assigned to Bayer AG, 16 October 1991.

Whitfill, D. L., Pober, K. W., Carlson, T. R., Tare, U. A., Fisk, J. V., Billingsley, J. L., 2005. Method for drilling depleted sands with minimal drilling fluid loss. US Patent6 889 780, assigned to Halliburton Energy Services, Inc., Duncan, OK, 10 May 2005. < http://www.freepatentsonline.com/6889780.html >.

Wilcox, R. D., 1990. Surface area approach key to borehole stability. Oil Gas J. 88(9), 66 - 80. Yeager, R. R., Bailey, D. E., 1988. Diesel - based gel concentrate improves rocky mountain regionfracture treatments. In: Proceedings Volume. SPE Rocky Mountain Regional Management, Casper, Wyo, 11 - 13 May 1988, pp. 493 - 497.

Zhou, Z. J., Gunter, W. D., Jonasson, R. G., 1995. Controlling formation damage using claystabilizers: A review. In: Proceedings Volume - 2. No. CIM 95 - 71. 46th Annual Cim. PetroleumSociety Techical Management, Banff, Canada 14 - 17 May 1995.

9 pH 值控制剂

pH 值的含义是指氢离子浓度指数的数值,表示溶液酸性式碱性程度的数值,即所含氢离子浓度的常用对数的负值。例如,氢离子的浓度为 10^{-12} mol/L,意味着 pH 值为 12。

利用平衡常数 K 表征也可以,有时也用 pK 值。

水的离解程度很低,可离解为质子,更准确地说是离解为氢离子和氢氧根离子,即:

$$H_2O \longrightarrow H^+ + OH^- \tag{9.1}$$

在水中,氢离子是不稳定的,可能会与另一个水分子相连:

$$H^+ + H_2O \longrightarrow H_3O^+ \tag{9.2}$$

为简单起见,我们通常只考虑 H^+。方程式(9.1)中水的离解反应平衡常数见 9.3。

$$K_{H_2O} = \frac{[H^+][OH^-]}{[H_2O]} \tag{9.3}$$

浓度的物理量纲为 mol/L,所以 K_{H_2O} 的物理量纲也是 mol/L。

9.1 缓冲理论

缓冲理论是物理化学中的基础理论,见 Chang 写的书(2000)。质子酸 A = B − H 第一步分解为质子 H^+ 和碱 B^-。

$$BH \longrightarrow H^+ + B^-$$

$$B^- + H_2O \longrightarrow BH + OH^- \tag{9.4}$$

式(9.4)中含有两个反应式,平衡常数如下。

酸常数:

$$K_A = \frac{[H^+][B^-]}{[BH]} \tag{9.5}$$

碱常数:

$$K_B = \frac{[OH^-][BH]}{[B^-][H_2O]} \tag{9.6}$$

根据式(9.4)两个独立方程,可得出水的离解方程式:

$$H_2O \longrightarrow H^+ + OH^- \tag{9.7}$$

因此

$$K_A K_B = K_{H_2O} \tag{9.8}$$

进一步得出水的平衡式：

$$[H_2O]_0 = [H_2O] + [H^+] \tag{9.9}$$

$[H_2O]_0$是未离解的水浓度，由电中和机理可得：

$$[H^+] = [OH^-] \tag{9.10}$$

水的平衡常数很小，所以我们可以近似认为$[H_2O]_0 \approx [H_2O]$。

式(9.11)被称为水的离子积，接近$10^{-14} mol^2/L$。

$$[H_2O]_0 K_{H_2O} = [H^+][OH^-] \tag{9.11}$$

将式(9.9)、式(9.10)代入式(9.3)中，可得：

$$K_{H_2O} = \frac{[H^+]^2}{[H_2O]_0 - [H^+]} \tag{9.12}$$

从式(9.11)或式(9.9)可以计算出纯水的pH值。

利用以上公式延伸到弱酸弱碱，根据式(9.4)可以计算出缓冲体系中的pH值以及滴定曲线。常见的水基缓冲液成分见表9.1和图9.1、图9.2。

表9.1 常见的缓冲溶液(Kolditz,1967)

反应	pK_A
$HCl \longrightarrow H^+ + Cl^-$	-7
$H_2SO_4 \longrightarrow H^+ + HSO_4^-$	-3
$HSO_4^- \longrightarrow SO_4^{2-} + H^+$	1.92
$H_2SO_3 \longrightarrow HSO_3^- + H^+$	1.92
$HF \longrightarrow F^- + H^+$	3.14
$CH_3COOH \longrightarrow CH_3COO^- + H^+$	4.75
$H_2S \longrightarrow HS^- + H^+$	6.92
$NH_4^+ \longrightarrow NH_3 + H^+$	9.25
$H_2O \longrightarrow H^+ + OH^-$	15.47
草酸/草酸氢盐	1.27
马来酸/马来酸氢盐	1.92
富马酸/富马酸氢盐	3.03
柠檬酸/柠檬酸氢盐	3.13
氨基磺酸/氨基磺酸盐	1.0

续表

反应	pK_A
甲酸/甲酸盐	3.8
醋酸/醋酸盐	4.7
磷酸二氢/磷酸氢盐	7.1
铵/氨	9.3
碳酸氢盐/碳酸盐	10.4
富马酸/氢延胡索酸酯	3.0
苯甲酸/苯甲酸盐	4.2

甲酸　富马酸　氨基磺酸

图 9.1　有机弱酸

乙酸　乙二酸　丙二酸

顺丁烯二酸　丁二酸

乙二酸　苯甲酸

苯甲酸　邻甲基苯甲酸　苯四甲酸

图 9.2　碳酸和二元碳酸

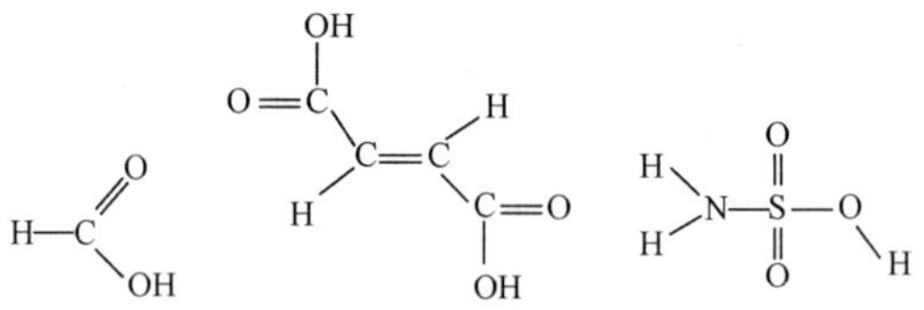

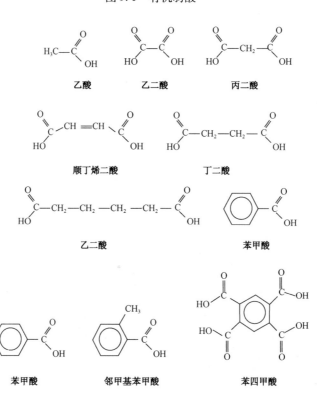

调整、维持 pH 值的缓冲液可以是弱酸盐或弱碱盐。如碳酸盐、碳酸氢盐、磷酸氢盐、甲酸、富马酸和氨基磺酸(Nimerick,1996)。

在压裂液中添加碳酸氢钠可提高各种冻胶的温度稳定性,维持 pH 值在 9.2~10.4 之间。

9.2 pH 值控制

缓冲剂的使用依据是 pH 值范围。对于压裂液来说,表 9.2 总结了几个缓冲体系。

表 9.2 缓冲体系(Putzig,2012)

pH 值范围	缓冲体系
5~7	富马酸
5~7	二乙酸钠
7~8.5	碳酸氢钠
9~12	碳酸钠
9~12	氢氧化钠

对于锆交联瓜尔胶体系,使用缓冲剂和 α - 羟基羧酸或 α - 羟基羧酸盐的混合物,可使交联延迟时间在较宽的 pH 值范围内可控。

参 考 文 献

Chang,R.,2000. Physical chemistry for the chemical and biological sciences. University ScienceBooks,Sausalito,California(Chapter 11).

Kolditz,L.(Ed.),1967. Anorganikum:Lehr – und Praktikumsbuch der anorganischen Chemie; miteiner Einführung in die physikalische Chemie. Dt. Verl. der Wiss.,Berlin,p. 413.

Moorhouse,R.,Matthews,L. E.,2004. Aqueous based zirconium(iv)crosslinked guar fracturingfluid and a method of making and use therefor. US Patent 6 737 386,assigned to Benchmark Research and Technology Inc.,Midland,TX,18 May 2004. < http://www. freepatentsonline. com/6737386. html >.

Nimerick,K.,1996. Fracturing fluid and method. GB Patent 2 291 907,assigned to Sofitech NV,7February 1996.

Putzig,D. E.,2012. Zirconium – based cross – linking composition for use with high pH polymersolutions. US Patent 8 247 356,assigned to Dorf Ketal Speciality Catalysts, LLC, Stafford, TX, 21 August 2012. < http://www. freepatentsonline. com/8247356. html >.

10 表面活性剂

表面活性剂多用于油田水处理剂,以提高水基液体的性能。压裂过后为了使地层内的碳氢化合物具有更高的导流能力,需要将压裂过后的地层表面形成水润湿表面。

烷基膦酸和磺化烷基膦酸容易吸附到固体表面上,特别容易吸附在地层中表面含烃的碳酸盐薄层表面。形成的吸附层非常薄,厚度相当于一个分子层,这明显比水润湿表面上的地层水层、水与表面活性剂的混合层更薄(Penny,1987a,1987b;Penny 和 Briscoe,1987)。

裂缝表面吸附的表面活性剂可以大大减少水和烃表面的润湿性,并提高水和碳氢化合物的界面张力。用碳氢化合物代替注入水,使碳氢化合物以较低毛细管张力通过孔隙进入裂缝通道,提高油气产量。

10.1 表面活性剂性能

用于砂岩储层增产处理的表面活性剂主要性能是降低液体体系的表面张力和接触角,从而控制液体滤失。然而,在应用过程中大量的表面活性剂会快速吸附在几英寸的砂岩地层表层。在这种情况下,其表面活性减少,并且不利于深度渗透。

Paktinat 等人于 2007 年对包括石油行业在内的实验室及油田现场应用的表面活性剂进行了总结描述。

为了研究几种不同表面活性剂的吸附性能,实验室内使用岩心柱对各种表面活性剂进行了吸附试验测试。此外,也采集了 Bradford,Balltown,Speechley 等砂岩地层的现场数据。将实验室和现场数据之间进行了数据比对。

相比于常规的表面活性剂,微乳液表面活性剂可提高地层的液体返排率。这一研究结果于 2007 年由 paktinat 等人提出用以减少地层伤害。

10.2 黏弹性表面活性剂

典型的黏弹性表面活性剂如 N-油酸-N,N-二(2-羟基乙基)-N-甲基氯化铵、油酸钾,与相应的活化剂如水杨酸钠溶液或氯化钾溶液混合时可形成黏弹性表面活性剂凝胶体系(Jones 和 Tustin,2007)。

甲基季铵化芥酸胺是在高温和高渗透油层压裂中应用的较为有效的黏弹性表面活性剂(Gadberry 等,1999)。

黏弹性表面活性剂在应用过程中存在破胶后表面活性剂溶液与地层内碳氢化合物形成低黏度水包油乳液的问题。这一现象使得两相分离工作较为困难,也使得井眼清理工作更为复杂。形成的乳液与传统的油基溶剂中表面活性剂不相溶或溶解度较小(Jones 和 Tustin,2007)。

10.2.1 阳离子表面活性剂

一些以季铵盐和季膦盐为基础的阳离子表面活性剂是已知在水和烃中溶解性能较好的表

面活性剂,并且也常作为相转移催化剂(Starks 等,1994)。

然而,特定的阳离子表面活性剂,如在水中形成黏弹性的表面活性剂却难溶于碳氢化合物,这一特点可使用两相溶解系数 $K_{o,w}$ 来表示,计算得到的数据接近于零。

两相溶解系数为物质在两种互不相溶介质(如油和水)中的平衡浓度比值。

$$K_{o,w} = \frac{C_o}{C_w} \tag{10.1}$$

两相溶解系数可以通过各种分析技术测定(Sharaf 等,1986)。

例如,可以用循环伏安法测试表面活性剂的临界胶束浓度,胶束的自扩散系数,及胶束的两相溶解系数(Mandal 和 Nair,1991)。同时,高效液相色谱法也是一种有效的测试技术(Terweij-Groen 等,1978)。

通常情况下,在烃类溶剂中溶解度高的阳离子表面活性剂是由多种长链烷基基团相互连接所组成的,可以通过十六烷基三丁基四氟离子和三辛基甲基铵离子的相互作用得以实现。

相反,用以形成黏弹性的阳离子表面活性剂通常是在每一个表面活性剂的端部接有一个直的长链烷烃(Jones 和 Tustin,2007)。

10.2.2 阴离子表面活性剂

一些阴离子表面活性剂在烃类溶剂中溶解度高,而在水中溶解度低。具有代表性的阴离子表面活性剂是双(2-乙基己酯)磺基琥珀酸钠(Manoj 等,1996)。这一混合物在水中不会形成黏弹性表面活性剂。因此,盐的加入会产生沉淀。热力学研究表明表面活性剂的胶束化过程是一个吸热过程,因此它是主要熵的贡献者。

表面活性剂的侧链尺寸越小在烃中的溶解性能越强。这主要是由于表面活性剂分子中含有的侧链越小,则在烃类溶剂中越易形成逆胶束状态,利用这一性能,可制备烃中逆胶束促进剂(Jones 和 Tustin,2007)。

通过改变表面活性剂主链上支链的数量和类型,可以使得表面活性剂在特殊烃类溶剂中的溶解性能变得更好或更差。通常情况下侧链由 α-碳原子组成。将适当的侧链引入表面活性剂主链,可以提高表面活性剂的黏弹性和溶解度等综合性能。图 10.1 介绍了 β-支链脂肪酸的合成过程。

合成 2-甲基油酸甲酯(Jones 和 Tustin,2007):使用四氢呋喃和正庚烷清洗钠氢化物。在氮气环境下添加 1,3-二甲基-3,4,5,6-四氢-2(1H)-嘧啶酮,并搅拌混合均匀。然后在 2h 的周期内滴加甲基油酸盐,将混合好的反应物加热反应 12h(控制反应的倒流),然后冷却至 0℃。

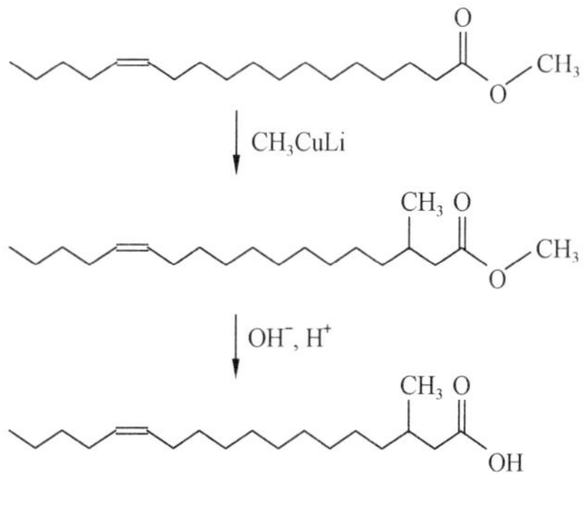

图 10.1 β-支链脂肪酸的合成过程
(Jones 和 Tustin,2007)

接下来继续反应,重新加热 2h 并在反应物中加入甲基碘化物,最后将反应物冷却至 0℃,并与水混合冷淬。将合成产物集中在真空中进行柱状色谱净化,最终得到类似于黄色油状物的 2-甲基油酸甲酯。

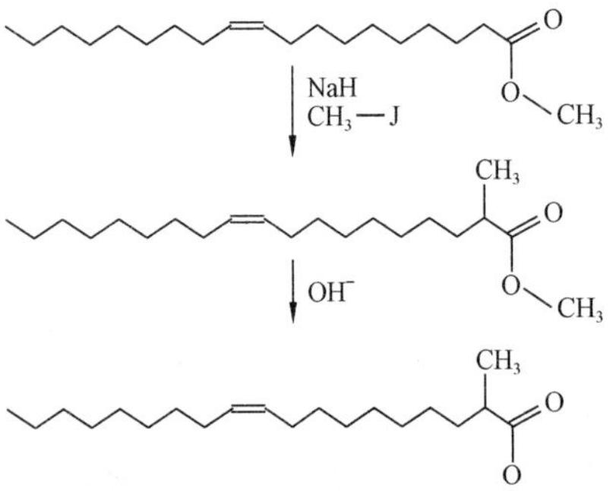

图10.2描述了2-甲基油酸甲酯的水解及酯化过程。当10%的2-甲基油酸钾与含16% KCl的盐溶液等体积混合后即可形成凝胶体系。

当凝胶与烃类物质,如庚烷接触后,其结果是凝胶体系黏度降低并形成两个低黏度液相:

(1)上部为油相;

(2)下部为分散较少的水相。

这一过程没有观察到乳液的生成。利用薄层色谱和红外光谱测试均显示了油酸在两相中分别存在。

由于内部胶束结构重新排列,破坏了表面活性剂形成的凝胶体系结构,其中被分解的油酸溶解在油相中。因此,油酸的破碎率比等效线性油酸更快。图10.3描述了室温条件下无分支油酸钾凝胶和支链甲基油酸钾凝胶强度与时间变化的关系。图中显示的两种凝胶制备方式相同,均为10%油酸溶液与16%氯化钾等体积混合,形成的凝胶体系混入等体积正庚烷中。

图10.2 2-甲基油酸甲酯的合成过程(Jones和Tustin,2007)

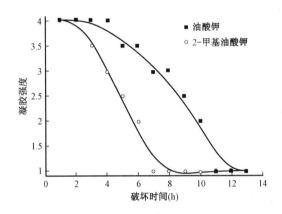

图10.3 凝胶强度与时间变化关系(Jones和Tustin,2007)

凝胶强度是用于表述表面活性剂凝胶流动性能的一种半定量测量值。对于表面活性剂凝胶强度主要分为4类,见表10.1。

表10.1 凝胶强度分类(Jones和Tustin,2010)

序号	强度
1	初始黏度
2	弱凝胶
3	舌形凝胶
4	变形非流动性凝胶

采用红外光谱分别测试2-甲基油酸钾凝胶破胶后的$K_{o,w}$为0.11,相比之下油酸钾的$K_{o,w}$值接近于0。

在破胶后,油酸类表面活性剂凝胶很少形成乳液或不形成乳液的这一性能更能满足油井压裂液现场技术需求。性能优异的压裂液流体不会伤害地层基质,因此避免形成乳液也有利于减少地层伤害(Jones 和 Tustin,2007)。

10.2.3 溴化阴离子表面活性剂

离子型表面活性剂的相互作用是复杂的,过量的反离子会减少表面活性剂端部带电荷基团之间的斥力作用。表面活性剂胶束端部分子相比于胶束内部分子承受更大的分子间作用力。这种作用力的不均匀分布促使柱状胶束长度不断增长。随后,柱状胶束之间相互缠绕最终形成黏弹性表面活性剂凝胶体系(Lee 等,2010)。

常规表面活性剂在使用过程中的一个关键问题是破胶后形成稳定水包稠油乳状液的可能性较大,这种乳液稳定性较强,难以分离并造成地层伤害。这种现象产生的原因主要是传统黏弹性表面活性剂在烃中的溶解度有限。

然而,研究表明传统表面活性剂通过改性,可以改善破胶后的乳化性能,如油酸钾,或将表面活性剂溴化(Lee 等,2010)。油酸的溴化反应是在正己烷溶液中加入 HBr 并添加乙酸溶液进行的,可以获得 9 - 溴硬脂酸。

表 10.2 中展示了不同浓度 9 - 溴硬脂酸与 8%氯化钾溶液的两相分配系数及破胶时间。从表 10.2 可以看出,表面活性剂的溴化会显著改变其两相溶解系数,但不会改变破胶时间。

表 10.2 两相溶解系数与破胶时间数据表(Lee 等,2010)

浓度(%)	两相溶解系数(%)		破胶时间(h)	
	BET	OLE	BET	OLE
1	17.1	0	12	14
5	13.28	0	82	84
10	13.49	0	94	98

注:BES 为 9 - 溴硬脂酸,OLE 为油酸。

此外,表面活性剂溴化后会保留其原有的黏弹性能,而这一性能是在石油行业应用过程中最重要的性能之一。在压裂过程中,压裂液进入地层后,剪切应力降低,使得表面活性剂体系黏弹性快速恢复。然而,溴化后的表面活性剂凝胶零剪切黏度降低了,其黏度值由原来的 528Pa·s 降低至 180Pa·s。相比之下,8%的氯化钾溶液配制的 5%溴化表面活性剂凝胶在剪切速率为 100/s 的黏度值也由 0.48Pa·s 降低到 0.16Pa·s。

参 考 文 献

Gadberry,J. F.,Hoey,M. D.,Franklin,R.,Del Carmen Vale,G.,Mozayeni,F.,1999. Surfactantsfor hydraulic fracturing compositions. US Patent 5 979 555,assigned to Akzo Nobel NV,9November 1999.

Jones,T. G. J.,Tustin,G. J.,2007. Surfactant comprising alkali metal salt of 2 - methyl oleic acidor 2 - ethyl oleic acid. US Patent 7 196 041,assigned to Schlumberger Technology Corp.,Ridgefield,CT,27 March 2007. < http://www. freepatentsonline. com/7196041. html >.

Jones,T. G. J.,Tustin,G. J.,2010. Process of hydraulic fracturing using a viscoelastic wellbore fluid. US Patent 7 655 604,assigned to Schlumberger Technology Corp.,Ridgefield,CT,2 February 2010. < http://www. freepatentsonline. com/7655604. html >.

Lee,L.,Salimon,J.,Yarmo,M. A.,Misran,M.,2010. Viscoelastic properties of anionic brominated surfactants. Sains

Malays. 39(5), 753–760.

Mandal, A. B., Nair, B. U., 1991. Cyclic voltammetric technique for the determination of thecritical micelle concentration of surfactants, self–diffusion coefficient of micelles, andpartition coefficient of an electrochemical probe. J. Phys. Chem. 95(22), 9008–9013. http://dx.doi.org/10.1021/j100175a106.

Manoj, K. M., Jayakumar, R., Rakshit, S. K., 1996. Physicochemical studies on reverse micelles ofsodiumbis(2–ethylhexyl)sulfosuccinate at low water content. Langmuir 12(17), 4068–4072. http://dx.doi.org/10.1021/la950279a.

Paktinat, J., Pinkhouse, J. A., Williams, C., Clark, G. A., Penny, G. S., 2007. Field case studies: damage preventions through leakoff control of fracturing fluids in marginal/low–pressure gasreservoirs. SPE Prod. Oper. 22(3), 357–367.

Penny, G. S., 1987a. Method of increasing hydrocarbon production from subterranean formations. US Patent 4 702 849, 27 October 1987.

Penny, G. S., 1987b. Method of increasing hydrocarbon productions from subterranean formations. EP Patent 234 910, 2 September 1987.

Penny, G. S., Briscoe, J. E., 1987. Method of increasing hydrocarbon production by remedial welltreatment. CA Patent 1 216 416, 13 January 1987.

Sharaf, M. A., Illman, D. L., Kowalski, B. R., 1986. Chemometrics. Wiley, New York.

Starks, C. M., Liotta, C. L., Halpern, M., 1994. Phase–transfer catalysis: fundamentals, applications, and industrial perspectives. Chapman & Hall, New York.

Terweij–Groen, C. P., Heemstra, S., Kraak, J. C., 1978. Distribution mechanism of ionizablesubstances in dynamic anion–exchange systems using cationic surfactants in high–performanceliquid chromatography. J. Chromatogr. A 161, 69–82. http://dx.doi.org/10.1016/S0021-9673(01)85213-4.

11 阻 垢 剂

在石油行业的某些操作中,如生产、增产措施和运输,存在有结垢的风险。当注入过程的温度发生变化时,过饱和溶液中就有可能出现结垢现象。

另外,如果两种会形成沉淀的化学物质聚集在一起,也会结垢,例如,氟化氢溶液遇到钙离子。从热力学的角度来看,存在一个稳定区域,一个亚稳定区域和一个不稳定区域,由双结点曲线和不稳定曲线分别区分开。

垢一般由碳酸钙、硫酸钡、石膏、硫酸锶、碳酸铁、铁的氧化物、硫化铁和镁盐组成(Keatch,1998)。对此已有专著进行论述,如腐蚀和结垢手册(Becker,1998),以及在文献中关于垢沉积的综述文章(Crabtree 等,1999)。还有针对北海碳酸盐岩储层(Jordan 等,2003,2005)和墨西哥湾(Jordan 等,2002)的案例研究。最近的研究集中在绿色系统(Frenier 和 Hill,2004;Hasson 等,2011)。

11.1 分类及机理

这一问题基本上类似于在洗衣机中阻垢。因此,使用类似的化学品防止水垢沉积。阻垢有两种方法:一是添加能够与潜在结垢物质反应的化学物质,使其达到热力学稳定区域;二是添加可以抑制晶体生长的化学物质。

传统的阻垢剂是亲水性的,即它们溶解在水中。在井下挤压的情况下,阻垢剂需要吸附在岩石表面,避免被冲刷掉,直到其发挥作用。然而,在岩石表面的吸附会改变系统的表面张力和润湿性。为了克服这些缺点,开发了油溶性阻垢剂。还有外敷涂层阻垢剂。

通常,阻垢剂与缓蚀剂组合使用,而不单独使用(Martin 等,2005)。阻垢剂可分为两大类,即:

(1)热力学阻垢剂;
(2)动力学阻垢剂。

在含盐水油藏中,防垢是确保连续生产的重要因素。由于垢和腐蚀的控制不好,井会过早报废(Kan 和 Tomson,2010)。Viloria 等提出了两种动力学阻垢剂的操作方法(2010):

(1)吸附效果;
(2)结晶点的形态变化。

由于吸附作用,阻垢剂分子占据了形成垢的分子优先成核点上。因此,晶体无法找到表面上黏附的活性位置,就无法成形晶核。

另一种抑制剂的机制是基于吸附机理,即形态学变化机制,在抑制剂存在下可以防止晶体的形成。根据抑制剂的特点和在基质上的性质,抑制剂可能会吸附在晶体网络上,形成复杂的表面或网络,使其难以在活性位置上保持或生长。

海水通常与近海油田地层水反应,生成钡、钙和硫酸锶沉淀,影响石油生产。在某些油田,碳酸钙是主要的问题。

在某些地区,地层水电化学变化很大(Duccini 等,1997)。例如,在北海中部,钡离子从几毫克每升到几克每升变化。此外,pH 值从 4.4 变化到 7.5。最高,测到过 pH 值 11.7。在北海南部地区,水是高盐度,并且富含硫酸盐和酸的化合物。理想的阻垢剂应具有以下性能(Duccini等,1997):

(1)低阻垢剂浓度时可以高效抑制垢的产生;
(2)与海水和地层水配伍;
(3)具有平衡吸附与解吸能力,可以使化学剂缓慢均匀地释放进入产出水中;
(4)高温稳定性;
(5)低毒性和高生物降解能力;
(6)低成本。

阻垢剂可以粗分为有机和无机两类(Viloria 等,2010)。无机阻垢剂包括缩合磷酸盐,如聚偏磷酸盐或磷酸盐。合适的有机阻垢剂是聚丙烯酸(PAA)、膦基羧酸、磺化聚合物和膦酸盐(Duccini 等,1997)。

膦酸盐在高温下效果好,而磺化聚合物在较低的温度下有效(Talbot 等,2009)。包含膦酸盐和磺酸盐基团的共聚物可以用于制备和增强用于较大温度范围的阻垢剂。已有证据显示,在水基体系中,顶端带有乙烯基磺酸/丙烯酸共聚物的膦酸盐可以有效阻止硫酸钡成垢(Talbot等,2009)。表 11.1 中给出了阻垢剂的基本问题。

表 11.1 阻垢剂类型(Viloria 等,2010)

阻垢剂类型	局限性
无机聚磷酸盐	遇水溶解并由于温度、pH 值、溶液质量和浓度、磷酸盐类型及某些酶的存在而生成磷酸钙沉淀
有机聚磷酸盐	一定温度下遇水溶解。钙离子浓度较高时无效,用量较大
羧酸聚合物	钙容忍度有限(2000mg/L),虽然有些可以用于浓度高于 5000mg/L 容溶液中,但需要较大的用量
乙二胺四乙酸	价格昂贵

11.1.1 热力学抑制剂

热力学抑制剂是络合剂和螯合剂,适用于特定的垢。例如,对硫酸钡垢,常见的化学品是乙二胺四乙酸(EDTA)和氨基三乙酸。碳酸钙的溶解性可通过改变 pH 值、二氧化碳(CO_2)分压来调整。溶解度随着 pH 值的降低和二氧化碳分压的升高而提高,且随着温度的升高而降低。

然而,通常溶解度随温度升高而升高。溶解度的温度系数有赖于溶解焓。溶解的放热焓,导致了随着温度的升高而溶解度下降;反之亦然。

11.1.2 动力学抑制剂

水解过程的动力学抑制剂也可以有效防止水垢沉积(Sikes 和 Wierzbicki,1996)。这可以理解为立体专一性和非特异性机制阻垢。

11.1.3 依附性抑制剂

另一种阻垢机理是基于粘连的抑制剂。某些化学物质只是简单地抑制晶体黏附于金属表面。这些都是表面活性剂。

11.1.4 螯合剂的干扰

已有证据表明,用于压裂液的微量金属螯合剂,会影响广泛使用的硫酸钡阻垢剂的性能。乙二胺四乙酸、柠檬酸和葡萄糖酸,会使某些阻垢剂,如膦酸酯、聚羧酸盐、磷酯,在低浓度为 0.1mg/L 时完全失效。这样会影响阻垢剂程序的确定和阻垢效果。

这个结论来自于模拟北海在 pH 值为 4 和 6 时的结垢系统实验。阻垢剂研究的浓度分别为 50mg/L 和 100mg/L。在 pH 值为 4 和 6 观察到的有机螯合剂有很大负面影响。唯一不受这些干扰因素影响的阻垢剂是聚乙烯基磺酸(PVS)(Barthorpe,1993)。

11.2 数学模型

已经建立了相应的数学模型(Shuler 和 Jenkins,1989;Mackay 等,1998,Mackay 和 Sorbie,1999,1998)。可以用热力学和电化学模型模拟碳酸铁和硫化铁的结垢过程(Malandrino 等,1998;Mackay 和 Sorbie,1998,2000;Anderko,Zhang 等,2000)。精确的模型可以预测 pH 值、结垢指数、密度以及对抑制剂的要求。实验数据已经验证了模型,并且给出了分析误差。因此,开发了相应的结垢预测软件(Kan 和 Tomson,2010)。

硫酸盐结垢的趋势,如硫酸钙、钡和天青石,以及岩盐的垢,与盐水 pH 值的相关性较弱。相反,碳酸盐,如方解石、白云石、菱铁矿以及硫化物垢是可以酸溶的。因此,其结垢倾向与盐水 pH 值显著相关。对于 pH 值敏感型的垢,结构预测更为复杂(Kan 和 Tomson,2010)。

11.2.1 最佳用量

已经有关于阻垢剂最佳用量计算方法的描述(Mikhailov 等,1987)。该方法首先关注水的化学成分和温度。从这些参数可以计算一个稳定性指数,从而预测阻垢剂的最佳用量。

在盐水中形成的碳酸钙($CaCO_3$)、硫酸钙($CaSO_4$)和硫酸钡($BaSO_4$)垢会影响渗透率。因此,在新的裂缝中使用阻垢剂可以防止在裂缝中形成水垢。文献中有关于水力压裂液配方中使用阻垢剂的描述(Watkins 等,1993)。

11.2.2 沉降挤压法

在沉降挤压法中,阻垢剂在地层的岩石孔隙中反应形成不溶性盐沉淀。例如,磷酸盐阻垢剂和螯合钙可以进行沉降挤压处理。此外,膦羧酸盐也已用于沉降挤压处理。聚环氧琥珀酸对于挤压处理是有效的(Brown 和 Brock,1995)。

阴离子型阻垢剂和多价阳离子的盐溶解在碱性水溶液中,提供了一个在碱性条件下可以互溶的,既有阻垢阴离子,又有多价阳离子的解决方案。然而,在低 pH 值时抑制剂是不溶的。在溶液中加入一种化合物,可以减缓碱性溶液 pH 值降低的速率。可以通过配方调整溶液 pH 值降低的速度(Collins,2000)。

在近井的高压处理模型中,假设井周围的流场是径向的。已经研究了近井地带是否为严格的非径向流动是影响高压处理的主要因素。人们已经发现,压裂井比非压裂井的高压有效周期更长。

此外,计算结果表明压裂井中吸附在裂缝表面的抑制剂本身对处理周期没有影响。在压裂井中,抑制剂与地层岩石的接触距离比径向处理的基质更远,接触也更缓慢(Rakhimov 等,2010)。

11.3 抑制剂化学

根据化学分类,抑制剂可以粗略地划分为酸用抑制剂和络合剂。近期文献中描述的阻垢剂见表11.2。

表11.2 阻垢剂

化合物	参考文献
1-羟基亚乙基-1,1-二磷酸	He 等(1999)
碳酰肼,$H_2N-NH-CO-NH-NH_2$	Mouche 和 Smyk(1995)
胺烷基膦酸和羧甲基纤维素或聚(丙烯酰胺)	Kochnev 等(1993)
聚丙烯酸和铬	Yan(1993)
聚丙烯酸盐①	Watkins 等(1993)
氨基亚甲基膦酸②	Graham 等(2000)
膦酰基甲基化的聚(胺)	Singleton 等(2000)
磺化聚丙烯酸共聚物	Chilcott 等(2000)
双[四(羟甲基)鏻]硫酸盐	Larsen 等(2000)
膦酸盐	Holzner 等(2000)和 Jordan 等(1997)
羧甲基菊粉	Kuzee 和 Raaijmakers(1999)
聚羧酸盐	Dobbs 和 Brown(1999)
米糠提取物膦酸酯类	Zeng 和 Fu(1998)
聚膦基马来酸酐	Yang 和 Song(1998)
N,N-己二烯-N-烷基-N-(磺基烷基)铵甜菜碱共聚物(带有 N-乙烯基吡咯烷酮或丙烯酰胺),二烯丙基二醇牛磺酸盐酸盐($CH_2=CH-CH_2Cl \times CH_3-NH-CH_2-CH_2-SO_3^- Na^+$)	Fong 等(2001)
氨基三亚甲基膦酸	Kowalski 和 Pike(2001)和 Tantayakom 等(2004,2005)
亚甲基膦酸	Tantayakom 等(2005)

① 在硼交联压裂液中应用。
② 高温条件。

11.3.1 水溶性抑制剂

11.3.1.1 酸

盐酸和氢氟酸等无机酸以及甲酸等有机酸,都可以用来降低 pH 值。酸通常与表面活性剂一起使用。

用作阻垢剂的酸具有强腐蚀性。其有效性已经在实验室中进行了测试。参数包括酸的类型、金属溶解性能、温度、缓蚀剂的类型和浓度、酸与金属接触的时间以及其他化学添加剂的效果(Burger 和 Chesnut,1992)。可以通过酸处理除去硫化铅和硫化锌水垢沉淀(Jordan 等,2000)。

11.3.1.2 氢氟酸

通过注入含有 HF 的酸液是可以改进渗透率伤害的。这种方法可以提高钙质和硅质地层的油气产量。

大多数砂岩地层是由 70% 以上的石英砂（即二氧化硅），与不同量的碳酸盐岩、白云岩和硅酸盐等胶结材料黏合在一起构成的。硅酸盐包括黏土和长石。常用的砂岩地层处理方法是从井筒注入氢氟酸，并使其与近井地层发生反应。

氢氟酸显示出与硅质矿物的高反应性能，如黏土和石英砂。例如，氢氟酸与蒙皂石、高岭石、伊利石和绿泥石等自生黏土反应非常迅速，特别是在温度高于 65℃ 时。同样地，氢氟酸能够溶解硅质矿物。

氢氟酸与地层中的钠、钾、钙和镁等金属离子接触，可能会发生不良的沉淀反应。

砂岩或硅质岩层以及灰岩地层可以用水基处理方式进行处理，这种方式含有氢氟酸源化合物与硼化合物和膦酸、酯或盐。这样的混合物通过抑制作用或阻止地层中不希望的垢的生成，从而增加地层的渗透率，这些垢包括氟化钙、氟化镁、氟硅酸钾、氟硅酸钠或氟铝酸盐。从而增加或提高了该地层的产量（Ke 和 Qu，2010）。

11.3.1.3 胶囊包裹的阻垢剂

这类阻垢剂可以在较长时间内化学释放（Powell 等，1995b；Hsu 等，2000）。微胶囊的配方包括囊衣材料及多种用途的混合材料，如 Kowalski 和 Pike（1999，2001）：

(1) 阻垢剂；
(2) 缓蚀剂；
(3) 杀菌剂；
(4) 硫化氢清除剂；
(5) 破乳剂；
(6) 黏土稳定剂。

11.3.1.4 螯合剂

微量的螯合剂，如 EDTA、柠檬酸或葡萄糖酸，可以降低阻垢剂的效率（Barthorpe，1993）。钙离子和镁离子浓度可以抑制硫酸钡的生成（Boak 等，1999）。研究了五元磷酸盐、六元膦酸、膦基聚（羧酸）（PPCA）盐和 PVS 阻垢剂。表 11.3 中给出用于胶囊包覆的稳定的链状螯合剂（Kowalski 和 Pike，1999）。图 11.1 是某些基于氨基酸的螯合剂的示意图。

表 11.3 用于包覆稳定的螯合剂（Kowalski 和 Pike，1999）

螯合剂	缩写
N-(3-羟丙基)亚氨基-N,N-二乙酸	3-HPIDA
N-(2-羟丙基)亚氨基-N,N-二乙酸	2-HPIDA
N-甘油亚氨基-N,N-二乙酸	GLIDA
二羟基异亚氨基-N,N-二乙酸	DHPIDA
甲基亚氨基-N,N-二乙酸	MIDA

续表

螯合剂	缩写
2-甲氧基乙基亚氨基-N,N-二乙酸	MEIDA
氨基亚氨基二乙酸（=氨基氨三乙酸钠）	SAND
乙酰氨基亚氨基二乙酸	AIDA
3-甲氧基丙基亚氨基-N,N-二乙酸	MEPIDA
三（羟甲基）甲基亚氨基-N,N-二乙酸	TRIDA

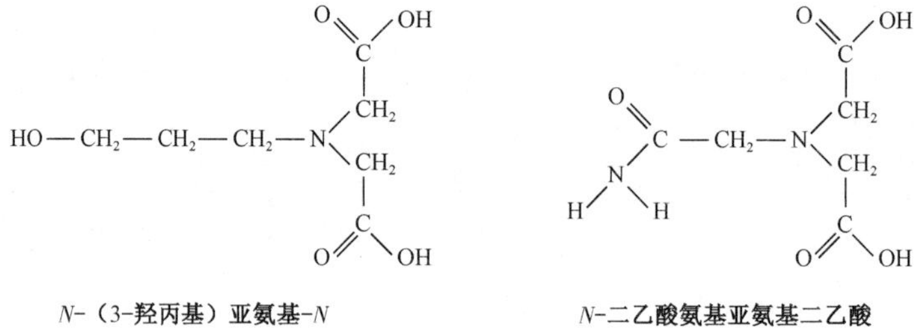

图 11.1 螯合剂

11.3.1.5 EDTA

碳酸钾、氢氧化钾和 EDTA 钾盐浓缩溶液是常规的重晶石垢的除垢剂。碳酸盐垢可以用简单的无机酸（如盐酸）溶解（Jones 等，2008）。此外，可以用表面活性剂控制流体的黏度。N-二十二烯-N,N-双（2-羟乙基）-N-甲基氯化铵可以作为表面活性剂使用（Jones，2008）。

与卤水混合时，这些表面活性剂能形成蠕虫状胶束。胶束的结构使流体具有黏弹性，当流体接触碳氢化合物，很快失去黏弹性，使胶束结构变化或解体。

与碳氢化合物及水接触后，表现出的不同黏度，可以用来选择性地进行垢处理。因此，可以优先从含油气区清除垢。这样，可以增加油气产量，而不会不大幅增加含水（Jones 等，2008）。

通过适当的工艺，EDTA 可以再生。方程（11.1）是硫酸钡垢溶解及后续分离，以及 EDTA 再生的简化方程（Keatch，2008）。

$$EDTA-K_4 + K_2CO_3 + BaSO_4 \longrightarrow EDTA-K_2Ba + K_2CO_3 + K_2SO_4$$
$$K_2CO_3 + 2HCl \longrightarrow 2KCl + H_2O + CO_2$$
$$EDTA-K_2Ba + K_2SO_4 \longrightarrow EDTA-K_4 + BaSO_4 \downarrow \tag{11.1}$$

11.3.1.6 膦酸酯

早先的研究已表明，五元膦酸酯和六元膦酸酯等氨基膦酸型抑制剂的热稳定性不如 PVS 和 S-Co 等聚合物。因此，磷酸基的材料不适用于高温油藏系统。

然而，最近的基于不同氨基膦酸抑制剂的研究显示，某些材料的耐温可以超过 160℃（Graham 等，2002）。

膦酸基系列阻垢剂在 160℃ 进行热老化后,阻垢剂可以防止动态试验中碳酸盐结垢。不过部分膦酸酯化合物在老化后阻止硫酸盐结垢的性能降低了(Dyer 等,2004)。

酯化膦酰基或膦酸与长链醇配合是有效的油溶性阻垢剂,如作为蜡或沥青质的抑制剂或油田生产的分散剂。相应的酯类是膦酸与醇共沸缩合反应制备得到的。另一种酯的制备是不饱和羧酸酯与亚磷酸盐或次磷酸盐调聚剂调聚反应制备得到的(Woodward 等,2004)。

相比之下,实验室研究表明,将磷酸酯膦酸换为乙烯砜共聚物类阻垢剂,可以延长阻垢时间(Jordan 等,2005)。

在压裂中广泛使用固体包裹型阻垢剂(钙和镁的聚磷酸盐)(Powell 等,1995,1996)。这种抑制剂因为有外敷层,可以与硼交联和锆交联的压裂液以及泡沫压裂液配伍。

这种涂层对释放速率有短期效应。固体衍生物的化学组成可以延长释放时间。

11.3.1.7 碱金属硫酸盐

在溶解重晶石的研究中,可以使用 EDTA 基和二乙三胺五乙酸基螯合剂,已经证实,加入草酸根离子等二羧酸添加剂可以改善螯合剂的性能。然而,其他添加剂如丙二酸盐和琥珀酸盐会降低其有效性。

草酸根离子催化了螯合剂和重晶石之间形成两个配位体的表面络合反应。由于空间位阻效应,阻止了两个配位体发生表面络合反应形成表面络合物,其他二元羧酸没有这种现象。

进一步研究天青石($SrSO_4$)、石膏($CaSO_4 \cdot 2H_2O$)和硬石膏($CaSO_4$)等其他类型重晶石垢,可以发现优化的垢溶解剂仅对某一种重晶石垢有效,但对其他类型的垢一般无效(Mendoza 等,2002)。

11.3.1.8 可生物降解的阻垢剂

许多石油公司要求使用环保型压裂液。压裂液是由多种具有特定功能的化合物组成的。压裂液也含有阻垢剂。在这里,我们不讨论压裂液的基本问题,这是第 1 章的内容。可生物降解的螯合剂是要讨论的内容(Crews,2006)。

11.3.1.9 亚氨基二琥珀酸钠

该化合物是一种马来酸衍生物。其主要用作二价和三价离子的螯合剂。这些配合物离子会导致乳化,从而结垢,可以改变酶破胶剂的性能,并导致交联冻胶不稳定,因此该螯合剂可以避免这些离子出现不良影响。

11.3.1.10 羟乙基亚氨基二乙酸二钠

这是少数氨基羧酸螯合剂之一,很容易生物降解。有利于易于结垢、降解酶、使交联冻胶不稳定的二价和三价离子的螯合作用。

11.3.1.11 葡萄糖酸钠和葡庚糖酸钠

这些多元醇常用于钙、镁、铁、锰和铜等矿物的螯合。也可用于复杂的延迟交联的钛酸盐、锆酸盐和硼酸盐离子的螯合。还是确保酶破胶剂和交联冻胶稳定性的优良的铁离子络合剂。

11.3.1.12 聚天冬氨酸钠

该化合物也被称为聚天冬氨酸。可作为多种类型二价和三价离子的螯合物。常用于破乳和阻垢。

聚天冬氨酸基的化学品已被确定为环保及可生物降解的油田化学品。可以在油藏注盐水提高采收率的作业中用作缓蚀阻垢剂。具有良好的钙络合性能。在 pH 值为 5 时,聚天冬氨酸的耐钙离子的浓度可以达到 8500~7500mg/L,相比之下,有机膦酸和马来酸聚合物产品的耐钙离子浓度仅有 5000mg/L。

浓度为 5% 的聚天冬氨酸,钙的络合性能性优于有机膦酸和马来酸的聚合物产品。同时,聚天冬氨酸还不影响油水的分离过程(Fan 等,2001)。

聚天冬氨酸化学品可能不仅仅是专门的阻垢剂,而且还可以在预处理中有助于其他阻垢剂的使用。聚天冬氨酸预处理溶液在低 pH 值时,有利于增加膦酸酯阻垢剂在岩石表面的吸附量(Montgomerie 等,2004)。

有人建议在或接近井场的生物反应器中合成聚天冬氨酸等处理剂。更直接的是引入能产生处理材料的井下嗜热古细菌或其他嗜热菌或生物直接在井下作业(Kotlar 和 Haugan,2005)。

11.3.2 油溶性阻垢剂

乙二胺四亚甲基膦酸或双六亚甲基三胺五(亚甲基膦酸)等膦酸类碱性化合物可以作为油溶性阻垢剂使用。其他合适的化合物是丙烯酸共聚物、PAA、PPCA 或膦酸酯。这些碱性化合物与胺类化合物混合形成油溶性混合物(Reizer 等,2002)。具有 12~16 个碳原子的叔烷基伯胺是油溶性的,并且是有效的油溶性阻垢剂。

11.3.2.1 芦荟类阻垢剂

芦荟阻垢剂是将芦荟胶溶解在水里。芦荟胶是由聚(糖)组成的,在 60℃ 和 90℃ 条件下,可溶解于水。在长链羧基、醇基团的存在条件下,可与 Ca^{2+} 和 Mg^{2+} 等二价离子相互作用。

与化学合成的抑制剂不同,芦荟植物胶的活性成分是天然化合物。这种阻垢剂可以应用在低和高的钙离子浓度条件下,并且没有在水解过程中出现沉淀的限制。相反,水解有利于离子在溶液中的相互作用,因此,其作为阻垢剂的效率,甚至会增加(Viloria 等,2010)。

反应趋向于使钙形成钙凝胶,依据鸡蛋盒模型将钙封装起来。钙离子的捕集机理如图 11.2 所示。在一般情况下,多价离子与聚合物的相互作用形成了凝胶。这种现象也被称为物理交联。

凝胶的链与 Ca^{2+} 相互作用,聚积在一起。在系统受力或其他条件下,可以使系统趋于稳定,否则会恢复到原来的凝胶状态。

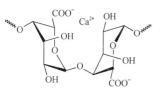

图 11.2 鸡蛋箱模型(Viloria 等,2010)

该模型假定钙离子作为一个桥梁,在两个相邻的不同的链上的两个羧基间形成离子键。根据这个聚(糖)模型,链与 Ca^{2+} 相互作用形成协同结构(a structure coordinated packaging)。

11.3.2.2 丙烯醛共聚物

在石油开发中,可以在水力压裂过程中注入丙烯醛和乙烯共聚物,降低硫化物垢的形成。而且,还可以增强防腐性能(Anon,2010)。

在生产过程中,丙烯醛保持一定下限浓度,在 24h 后丙烯醛的浓度基本上是稳定的,然后在接下来的 6 个月内逐渐降低。

11.3.3　高温油藏

常规的聚合物和有机膦酸阻垢剂不适用于高温高压储层。仅有少量市售的作为油田阻垢剂的化学材料可以用于150℃以上的油藏温度。

这些化学材料是乙烯基磺酸盐均聚物以及丙烯酸(AA)与乙烯基磺酸钠共聚物。其他聚合物,如聚马来酸、聚衣康酸和马来酸/丙烯酸共聚物,具有类似的热稳定性(Collins,1995)。已经有人进行了相关热稳定性试验,评价了pH值、离子强度和氧对传统聚合物和有机膦酸阻垢剂的影响,如,膦基聚羧酸、PVS、五元膦酸盐和六元膦酸盐(Graham等,1997,1998a,1998b;Dyer等,1999)。

正如上文所指出的,一般认为磷酸阻垢剂不能用于高温条件,最近研究表明,在严格的无氧和NaCl盐水条件下,膦酸盐抑制剂可以用于200℃油藏(Fan等,2010)。相反,膦酸盐阻垢剂在高温盐水中可能与Ca^{2+}沉淀。

文献中的商品名称

商品名	描述	供应商
Dequest® 2060	二乙烯三胺五亚甲基膦酸(Collins,2000)	Monsanto
Empol™(系列)	低聚油酸(Jones等,2008)	Henkel
Rhodafac® RS-410	聚氧基-1,2-亚乙基十三烷基羟基磷酸(Martin等,2005)	Rhodia
Scaletreat® XL14FD	聚马来酸盐(Collins,2000)	TR Oil Services Ltd.

参 考 文 献

Anderko, A. ,2000. Simulation of $FeCO_3$/FeS scale formation using thermodynamic and electrochemical models. In: Proceedings Volume. NACE International Corrosion Conference, Corrosion 2000, Orlando, FL, 26 – 31 March 2000.

Anon, 2010. Process for reducing iron sulfide scales and preventing corrosion during the explorationfor and production of hydrocarbons. IP. com J. 10(12B),22.

Barthorpe, R. T. ,1993. The impairment of scale inhibitor function by commonly used organicanions. In: Proceedings Volume. SPE International Symposium on Oilfield Chemistry, NewOrleans,2 – 5 March 1993,pp. 69 – 76.

Becker, J. R. ,1998. Corrosion and Scale Handbook. Pennwell Publishing Co, Tulsa.

Boak, L. S. , Graham, G. M. , Sorbie, K. S. ,1999. The influence of divalent cations on the performanceof $BaSO_4$ scale inhibitor species. In: Proceedings Volume. SPE International Symposium on Oilfield Chemistry, Houston, 16 – 19 February 1999, pp. 643 – 648.

Brown, J. M. , Brock, G. F. ,1995. Method of inhibiting reservoir scale. US Patent 5 409 062, assignedto Betz Laboratories, Inc. , Trevose, PA, 25 April 1995. < http://www. freepatentsonline. com/5409062. html >.

Burger, E. D. , Chesnut, G. R. ,1992. Screening corrosion inhibitors used in acids for downhole scaleremoval. Mater. Perf. 31(7),40 – 44.

Chilcott, N. P. , Phillips, D. A. , Sanders, M. G. , Collins, I. R. , Gyani, A. ,2000. The developmentand application of an accurate assay technique for sulphonated polyacrylate co – polymeroilfield scale inhibitors. In: Proceedings Volume. 2nd Annual SPE Oilfield Scale InternationalSymposium, Aberdeen, Scotland, 26 – 27 January 2000.

Collins, I. R. ,1995. Scale inhibition at high reservoir temperatures. In: Proceedings Volume. IBC Technical Services Ltd. Advances in Solving Oilfield Scaling International Conference, Aberdeen, Scotland, 20 – 21 November 1995.

Collins, I. R. ,2000. Oil and gas field chemicals. US Patent 6 148 913, assigned to BP ChemicalsLtd. , London, GB, 21

November 2000. < http://www. freepatentsonline. com/6148913. html >.

Crabtree, M. , Eslinger, D. , Fletcher, P. , Miller, M. , Johnson, A. , King, G. , 1999. Fighting scale—removal and prevention. Oilfield Rev. 11(3),30 – 45.

Crews, J. B. ,2006. Biodegradable chelant compositions for fracturing fluid. US Patent 7 078 370, assigned to Baker Hughes Inc. , Houston, TX,18 July 2006. < http://www. freepatentsonline. com/7078370. html >.

Dobbs, J. B. , Brown, J. M. ,1999. An environmentally friendly scale inhibitor. In: Proceedings Volume. NACE International Corrosion Conference, Corrosion 99, San Antonio,25 – 30 April1999.

Duccini, Y. , Dufour, A. , Harm, W. M. , Sanders, T. W. , Weinstein, B. , 1997. High performance oilfield scale inhibitors. In: Corrosion97. NACE International, New Orleans, LA. < http://www. onepetro. org/mslib/app/Preview. do? paperNumber = NACE – 97169\&societyCode = NACE >.

Dyer, S. J. , Graham, G. M. , Sorbie, K. S. ,1999. Factors affecting the thermal stability of conventional scale inhibitors for application in high pressure/high temperature reservoirs. In: Proceedings Volume. SPE International Symposium on Oilfield Chemistry, Houston,16 – 19 February 1999, pp. 167 – 177.

Dyer, S. J. , Anderson, C. E. , Graham, G. M. ,2004. Thermal stability of amine methyl phosphonate scale inhibitors. J. Petrol. Sci. Eng. 43,259 – 270. http://dx. doi. org/10. 1016/j. petrol. 2004. 02. 018.

Fan, J. C. , Fan, L. D. G. , Liu, Q. W. , Reyes, H. ,2001. Thermal polyaspartates as dual functioncorrosion andmineral scale inhibitors. Polym. Mater. Sci. Eng. 84, 426 – 427. < http://www. onepetro. org/mslib/servlet/onepetropreview? id = 00065005 >.

Fan, C. , Kan, A. T. , Zhang, P. , Lu, H. , Work, S. , Yu, J. , Tomson, M. B. ,2010. Scale prediction and inhibition for unconventional oil and gas production. In: SPE International Conferenceon Oilfield Scale. Society of Petroleum Engineers, Aberdeen, UK. < http://www. onepetro. org/mslib/app/Preview. do? paperNumber = SPE – 130690 – MS\ &societyCode = SPE >.

Fong, D. W. , Marth, C. F. , Davis, R. V. ,2001. Sulfobetaine – containing polymers and their utility ascalcium carbonate scale inhibitors. US Patent 6 225 430, assigned to Nalco Chemical Co. ,1 May 2001.

Frenier, W. W. , Hill, D. G. ,2004. Green inhibitors—development and applications for aqueous systems. In: Proceedings Volume. Volume 3 of Reviews on Corrosion Inhibitor Science and Technology. Corrosion – 2004 Symposium, New Orleans, LA, United States,28 March – 1 April,2004, pp. 6/1 – 6/39.

Graham, G. M. , Jordan, M. M. , Sorbie, K. S. , Bunney, J. , Graham, G. C. , Sablerolle, W. , Hill, P. ,1997. The implication of HP/HT(high pressure/high temperature) reservoir conditions onthe selection and application of conventional scale inhibitors: Thermal stability studies. In: Proceedings Volume. SPE Oilfield International Symposium on Chemistry, Houston,18 – 21 February 1997, pp. 627 – 640.

Graham, G. M. , Dyer, S. J. , Sorbie, K. S. , Sablerolle, W. , Graham, G. C. ,1998a. Practical solutions toscaling in HP/HT(high pressure/high temperature) and high salinity reservoirs. In: Proceedings Volume. 4TH IBC UK Conf. Ltd Advances in Solving Oilfield Scaling InternationalConference, Aberdeen, Scotland,28 – 29 January 1998.

Graham, G. M. , Dyer, S. J. , Sorbie, K. S. , Sablerolle, W. R. , Shone, P. , Frigo, D. ,1998b. Scale inhibitor selection for continuous and downhole squeeze application in HP/HT(high pressure/hightemperature) conditions. In: Proceedings Volume. Annual SPE Technical Conference, New Orleans,27 – 30 September 1998, pp. 645 – 659.

Graham, G. M. , Dyer, S. J. , Shone, P. , 2000. Potential application of amine methylene phosphonatebased inhibitor species in HP/HT(high pressure/high temperature) environments for improvedcarbonate scale inhibitor performance. In: Proceedings Volume. 2nd Annual SPE Oilfield Scale International Symposium, Aberdeen, Scotland,26 – 27 January 2000.

Graham, G. M. , Dyer, S. J. , Shone, P. ,2002. Potential application of amine methylene phosphonate – based inhibitor species in hp/ht environments for improved carbonate scale inhibitor performance. SPE Prod. Facil. 17,212 – 220. < http://www. onepetro. org/mslib/servlet/onepetropreview? id = 00060217 >.

Hasson, D., Shemer, H., Sher, A., 2011. State of the art of friendly green scale control inhibitors: areviewarticle. Ind. Eng. Chem. Res. 50(12), 7601 – 7607. http://dx.doi.org/10.1021/ie200370v.

He, S., Kan, A. T., Tomson, M. B., 1999. Inhibition of calcium carbonate precipitation inNaCl brinesfrom 25 to 90℃. Appl. Geochem. 14(1), 17 – 25.

Holzner, C., Kleinstueck, R., Spaniol, A., 2000. Phosphonate – containing mixtures (Phosphonathaltige Mischungen). WO Patent 0 032 610, assigned to Bayer AG, 8 June 2000.

Hsu, J. F., Al – Zain, A. K., Raju, K. U., Henderson, A. P., 2000. Encapsulated scale inhibitor treatments experience in the ghawar field, saudi arabia. In: Proceedings Volume. 2nd Annual SPE Oilfield Scale International Symposium, Aberdeen, Scotland, 26 – 27 January 2000.

Jones, T. G. J., Tustin, G. J., Fletcher, P., Lee, J. C. – W., 2008. Scale dissolver fluid. US Patent 7 343 978, assigned to Schlumberger Technology Corp., Ridgefield, CT, 18 March 2008. < http://www.freepatentsonline.com/7343978.html >.

Jordan, M. M., Sorbie, K. S., Chen, P., Armitage, P., Hammond, P., Taylor, K., 1997. The designof polymer and phosphonate scale inhibitor precipitation treatments and the importance ofprecipitate solubility in extending squeeze lifetime. In: Proceedings Volume. SPE InternationalSymposium on Oilfield Chemistry, Houston, 18 – 21 February 1997, pp. 641 – 651.

Jordan, M. M., Sjursaether, K., Bruce, R., Edgerton, M. C., 2000. Inhibition of lead and zincsulphide scale deposits formed during production from high temperature oil and condensatereservoirs. In: Proceedings Volume. SPE Asia Pacific Oil & Gas Conference, Brisbane, Australia, 16 – 18 October 2000.

Jordan, M. M., Sjuraether, K., Collins, I. R., Feasey, N. D., Emmons, D., 2002. Life cycle managementof scale control within subsea fields and its impact on flow assurance gulf of mexicoand the North Sea basin. Spec. Publ. R. Soc. Lond. 280, 223 – 253. < http://www.onepetro.org/mslib/servlet/onepetropreview?id=00071557 >.

Jordan, M. M., Kemp, S., Sorhaug, E., Sjursaether, K., Freer, B., 2003. Effective management ofscaling from and within carbonate oil reservoirs, North Sea basin. Chem. Eng. Res. Des. 81, 359 – 372. http://dx.doi.org/10.1205/02638760360596919.

Jordan, M. M., Sjursaether, K., Collins, I. R., 2005. Scale control within the North Seachalk/limestone reservoirs – the challenge of understanding and optimizing chemical – placement methods and retention mechanisms: Laboratory to field. SPE Prod. Facil. 20, 262 – 273. < http://www.onepetro.org/mslib/servlet/onepetropreview?id=SPE – 86476 – PA >.

Kan, A. T., Tomson, M. B., 2010. Scale prediction for oil and gas production. In: International Oiland Gas Conference and Exhibition in China. Society of Petroleum Engineers, Beijing, China. < http://www.onepetro.org/mslib/app/Preview.do?paperNumber=SPE – 132237 – MS\&societyCode=SPE >.

Ke, M., Qu, Q., 2010. Method for controlling inorganic fluoride scales. US Patent 7 781 381, assigned to BJ Services Company LLC, Houston, TX, 24 August 2010. < http://www.freepatentsonline.com/7781381.html >.

Keatch, R. W., 1998. Removal of sulphate scale from surface. GBPatent 2 314 865, 14 January 1998.

Keatch, R., 2008. Method for dissolving oilfield scale. US Patent 7 470 330, assigned to M – 1Production Chemicals UK Ltd., Aberdeen, GB, Oilfield Mineral Solutions Ltd., Edinburgh, GB, 30 December 2008. < http://www.freepatentsonline.com/7470330.html >.

Kochnev, E. E., Merentsova, G. I., Andreeva, T. L., Ershov, V. A., 1993. Inhibitor solution to avoidinorganic salts deposition in oil drilling operations—contains water, carboxymethylcelluloseor polyacrylamide and polyaminealkyl phosphonic acid and has improved distribution uniformity. SU Patent 1 787 996, assigned to Siberian Research Institute of the Oil Industry, 15January 1993.

Kotlar, H. K., Haugan, J. A., 2005. Genetically engineered well treatment microorganisms. GBPatent 2 413 797, assigned to Statoil Asa, 9 November 2005.

Kowalski, T. C. , Pike, R. W. , 1999. Microencapsulated oil field chemicals. US Patent 5 922 652, 13 July 1999.

Kowalski, T. C. , Pike, R. W. , 2001. Microencapsulated oil field chemicals. US Patent 6 326 335, assigned to Corsicana Technologies Inc. , 4 December 2001.

Kuzee, H. C. , Raaijmakers, H. W. C. , 1999. Method for preventing deposits in oil extraction. WO Patent 9 964 716, assigned to Cooperatie Cosun Ua, 16 December 1999.

Larsen, J. , Sanders, P. F. , Talbot, R. E. , 2000. Experience with the use of tetrakishydroxymethylphosphonium sulfate (THPS) for the control of downhole hydrogen sulfide. In: Proceedings Volume. NACE International Corrosion Conference, Corrosion 2000, Orlando, FL, 26 – 31 March 2000.

Mackay, E. J. , Sorbie, K. S. , 1998. Modelling scale inhibitor squeeze treatments in high crossflowhorizontal wells. J. Can. Pet. Technol. 39(10), 47 – 51.

Mackay, E. J. , Sorbie, K. S. , 1999. An evaluation of simulation techniques for modelling squeezetreatments. In: Proceedings Volume. Annual SPE Technical Conference, Houston, 3 – 6 October 1999, pp. 373 – 387.

Mackay, E. J. , Sorbie, K. S. , Jordan, M. M. , Matharu, A. P. , Tomlins, R. , 1998. Modelling of scaleinhibitor treatments in horizontal wells: Application to the alba field. In: Proceedings Volume. SPE International Symposium on Formation Damage Control, Lafayette, LA, 18 – 19 February 1998, pp. 337 – 348.

Malandrino, A. , Andrei, M. , Gagliardi, F. , Lockhart, T. P. , 1998. A thermodynamic model for PPCA (phosphino – polycarboxylic acid) precipitation. In: Proceedings Volume. 4th IBC UKConf. Ltd Advances in Solving Oilfield Scaling International Conference, Aberdeen, Scotland, 28 – 29 January 1998.

Martin, R. L. , Brock, G. F. , Dobbs, J. B. , 2005. Corrosion inhibitors and methods of use. US Patent 6 866 797, assigned to BJ Services Co. , 15 March 2005. < http://www. freepatentsonline. com/6866797. html >.

Mendoza, A. , Graham, G. M. , Farquhar, M. L. , Sorbie, K. S. , 2002. Controlling factors of EDTAand DTPA based scale dissolvers against sulphate scale. Prog. Min. Oilfield Chem. 4, 41 – 58.

Mikhailov, S. A. , Khmeleva, E. P. , Moiseeva, E. V. , Sleta, T. M. , 1987. Determination of the optimaldose of salt deposition inhibitors. Neft Khoz 7, 43 – 45.

Montgomerie, H. T. R. , Chen, P. , Hagen, T. , Wat, R. M. S. , Selle, O. M. , Kotlar, H. K. , 2004. Methodof controlling scale formation. WO Patent 2 004 011 772, assigned to Champion TechnologyInc. , Statoil Asa, Montgomerie Harry Trenouth Rus, Chen Ping, Hagen Thomas, Wat RexMan Shing, Selle Olav Martin, and Kotlar Hans Kristian, 5 February 2004.

Mouche, R. J. , Smyk, E. B. , 1995. Noncorrosive scale inhibitor additive in geothermal wells. USPatent 5 403 493, assigned to Nalco Chemical Co. , 4 April 1995.

Powell, P. J. , Gdanski, R. D. , McCabe, M. A. , Buster, D. C. , 1995a. Controlled – release scaleinhibitor for use in fracturing treatments. In: Proceedings Volume. SPE Oilfield International Symposium on Chemistry, San Antonio, 14 – 17 February 95, pp. 571 – 579.

Powell, R. J. , Fischer, A. R. , Gdanski, R. D. , McCabe, M. A. , Pelley, S. D. , 1995b. Encapsulated scaleinhibitor for use in fracturing treatments. In: Proceedings Volume. Annual SPE TechnicalCon ference, Dallas, 22 – 25 October 1995, pp. 557 – 563.

Powell, R. J. , Fischer, A. R. , Gdanski, R. D. , McCabe, M. A. , Pelley, S. D. , 1996. Encapsulated scaleinhibitor for use in fracturing treatments. In: Proceedings Volume. SPE Permian Basin Oil & Gas Recovery Conference, Midland, TX, 27 – 29 March 1996, pp. 107 – 113.

Rakhimov, A. Z. , Vazquez, O. , Sorbie, K. S. , Mackay, E. J. , 2010. Impact of fluid distribution on scaleinhibitor squeeze treatments. In: SPE EUROPEC/EAGE Annual Conference and Exhibition. Society of Petroleum Engineers, Barcelona, Spain. < http://www. onepetro. org/mslib/app/Preview. do? paperNumber = SPE – 131724 – MS \ &societyCode = SPE >.

Reizer, J. M. , Rudel, M. G. , Sitz, C. D. , Wat, R. M. S. , Montgomerie, H. , 2002. Scale inhibitors. USPatent 6 379 612, assigned to Champion Technology Inc. ,30 April 2002.

Shuler, P. J. , Jenkins, W. H. , 1989. Prevention of downhole scale deposition in the ninian field. In: Proceedings Volume. Vol. 2. SPE Offshore Europe Conference, Aberdeen, Scotland, 5 – 8 September 1989.

Sikes, C. S. , Wierzbicki, A. , 1996. Stereospecific and nonspecific inhibition of mineral scale and ice formation. In: Proceedings Volume. 51st Annual NACE International Corrosion Conference, Corrosion 96, Denver, 24 – 29 March 1996.

Singleton, M. A. , Collins, J. A. , Poynton, N. , Formston, H. J. , 2000. Developments in phosphonomethylated polyamine (PMPA) scale inhibitor chemistry for severe $BaSO_4$ scaling conditions. In: Proceedings Volume. 2nd Annual SPE Oilfield Scale International Symposium, Aberdeen, Scotland, 26 – 27 January 2000.

Talbot, R. E. , Jones, C. R. , Hills, E. , 2009. Scale inhibition in water systems. US Patent 7 572 381, assigned to Rhodia UK Ltd. , Hertfordshire, GB, 11 August 2009. < http://www.freepatentsonline.com/7572381.html >.

Tantayakom, V. , Fogler, H. S. , de Moraes, F. F. , Bualuang, M. , Chavadej, S. , Malakul, P. , 2004. Study of Ca – ATMP precipitation in the presence of magnesium ion. Langmuir 20, 2220 – 2226. < http://pubs.acs.org/doi/abs/10.1021/la0358318 >.

Tantayakom, V. , Fogler, H. S. , Chavadej, S. , 2005. Scale inhibitor precipitation kinetics. In: Proceedings Volume. 7th World Congress of Chemical Engineering, Glasgow, United Kingdom, 10 – 14 July 2005, pp. 85704/1 – 85704/8.

Viloria, A. , Castillo, L. , Garcia, J. A. , Biomorgi, J. , 2010. Aloe derived scale inhibitor. US Patent 7 645 722, assigned to Intevep, S. A. , Caracas, VE, 12 January 2010. < http://www.freepatentsonline.com/7645722.html >.

Watkins, D. R. , Clemens, J. J. , Smith, J. C. , Sharma, S. N. , Edwards, H. G. , 1993. Use of scale inhibitors in hydraulic fracture fluids to prevent scale build – up. US Patent 5 224 543, assigned to Union Oil Co. , California, 6 July 1993.

Woodward, G. , Jones, C. R. , Davis, K. P. , 2004. Novel phosphonocarboxylic acid esters. WO Patent 2 004 002 994, assigned to Rhodia Consumer Specialities L, Woodward Gary, Jones Christopher Raymond, and Davis Keith Philip, 8 January 2004.

Yan, T. Y. , 1993. Process for inhibiting scale formation in subterranean formations. WO Patent 9 305 270, assigned to Mobil Oil Corp. , 18 March 1993.

Yang, L. , Song, B. , 1998. Phosphino maleic anhydride polymer as scale inhibitor for oil/gas field produced waters. Oilfield Chem. 15(2) ,137 – 140.

Zeng, Y. B. , Fu, S. B. , 1998. The inhibiting property of phosphoric acid esters of rice bran extract for barium sulfate scaling. Oilfield Chem. 15(4) ,333 – 335,365.

Zhang, H. , Mackay, E. J. , Sorbie, K. S. , Chen, P. , 2000. Non – equilibrium adsorption and precipitation of scale inhibitors: corefloods and mathematical modelling. In: Proceedings Volume. SPE International Oil & Gas Conference and Exhibition in China, Beijing, 7 – 10 November 2000, pp. 1 – 18.

12 发 泡 剂

泡沫流体多用于敏感性储层压裂作业中(Stacy 和 Weber,1995)。泡沫流体的主要性能包括发泡剂、可重复利用及剪切稳定性等,并且可以在较宽的温度范围内形成稳定的泡沫体系。一般泡沫压裂液即使在较高的温度下也具有较好的黏度(Bonekamp 等,1993)。

泡沫压裂液性能优于常规压裂液,其主要是由于泡沫压裂液内含有较少的水,因此较常规压裂液相比其滤失更少,更适用于水敏性地层。同时,泡沫压裂液在压裂施工后返排液处理方面需要的花费也更少。此外,压裂后由于井筒内压力降低,有利于气体泡沫体积膨胀,这一性能可以将压裂过程中进入地层的流体返排流回井筒内。

泡沫压裂液也可用于携带支撑剂,这样有利于裂缝闭合后在地层内形成具有较高导流能力的裂缝通道。各种支撑材料包括树脂涂层支撑剂或无涂层陶粒及石英砂、矾土、陶瓷材料及玻璃珠。最佳支撑剂加量范围是每加仑泡沫压裂中加入 1～10 lb 的支撑剂(Dahanayake 等,2008)。

泡沫压裂液中气体含量为质量分数。其泡沫质量 Q 值以百分比形式表示,如式(12.1):

$$Q = \frac{V_\mathrm{f} - V_\mathrm{l}}{V_\mathrm{f}} \times 100\% \tag{12.1}$$

式中,V_f 是泡沫总体积,V_l 是泡沫中液体体积。因此,70%气含量的泡沫压裂液中含70%质量的气体。最近,已研制出含95%气体的泡沫材料。只需要在水中添加2%的阴离子表面活性剂,这种类型的泡沫就可以形成性能稳定的泡沫体系(Harris 和 Heath,1996)。

表面活性剂可改变发泡能力。例如,通过改变三烷基胺聚氧乙烯醚表面活性剂环境的酸碱性可以改变其发泡性能。例如向其中增加氢离子降低环境 pH 值,就可以使其由发泡体系变成无泡沫体系。然而继续向其中添加氢氧根离子,提高环境 pH 值,就可以恢复体系的发泡性能。图 12.1 描述了低 pH 值条件下表面活性剂中氨基基团的季铵化。

此外,可可甜菜碱和 α-磺酸盐是一种较好的发泡剂(Pakulski 和 Hlidek,1992)。优选的两性表面活性剂是月桂胺和十四烷基氧化铵的混合物。图 12.2 显示了十二烷基甜菜碱分子式。泡沫压裂液是可循环利用的流体(Chatterji 等,2007),改变压裂液的 pH 值可以使体系消泡。

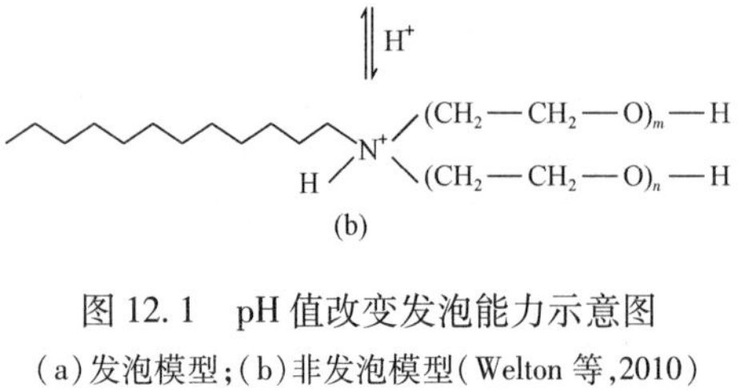

图 12.1　pH 值改变发泡能力示意图
(a)发泡模型;(b)非发泡模型(Welton 等,2010)

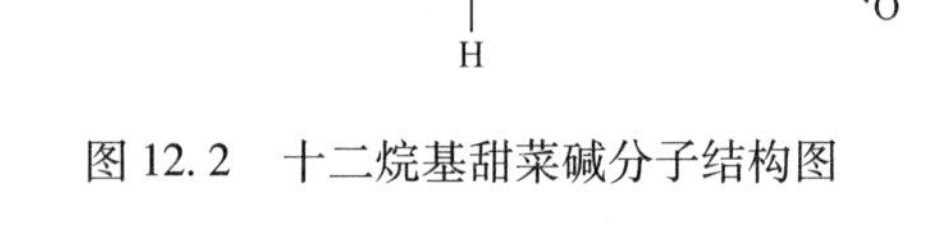

图 12.2　十二烷基甜菜碱分子结构图

12.1 环境安全型流体

传统的泡沫压裂液包括各种表面活性剂,通常含有发泡剂和稳泡剂,其主要作用是当气体与压裂液混合时促进泡沫产生并保证泡沫稳定性(Chatterji 等,2004)。

然而,单纯的发泡和稳泡性能已不能满足表面活性剂的环境要求。当表面活性剂进入地层后,不能完全降解的表面活性剂会污染环境水,破坏水质。

由蹄角粉水解制成的水解角蛋白发泡剂是一种无毒的环境友好型泡沫压裂液的添加剂。蹄角粉与石灰混合在高压釜加热反应可生成水解蛋白。这种蛋白质是一种市售水解蛋白,粉末状,有效含量为85%。

水解角蛋白粉末的非蛋白部分含有0.58%不溶性材料,其中还含有一些可溶性非蛋白物质,包括硫酸钙、硫酸镁和硫酸钾(Chatterji 等,2004)。

当需要进一步增加泡沫压裂液的黏度和泡沫稳定性时,需要在压裂液中添加额外的添加剂。常见的添加剂有碘、过氧化氢、硫酸铜和溴化锌。

12.2 液态二氧化碳泡沫

氮气泡沫可用于液态CO_2压裂(Gupta 等,2004)。发泡剂是一种非离子含氟醚类表面活性剂。压裂液中的液相为CO_2,气相为N_2。

氟化合物可以通过氟化反应直接合成制得,通过电化学氟化反应使含氟单体加成聚合以及含氟单体的氧化聚合。

含氟化合物由于缺少氯原子,从而不会消耗臭氧,这一性能使得化合物更加环保,不会破坏大气环境。通过优选,这种含氟类表面活性剂分子中至少有一种脂族氢原子。它具有较强的热力学稳定性及化学稳定性,并且更加环保,它的臭氧消耗能力几乎为零。六氟二甲苯就是这种化合物中的一种。

可用氟化醇烷基化方法制备含氟醚。其主要方法是由相应的含氟酰基类化合物或含氟酮化合物与无水含氟碱金属或无水氟化银在无水极性非质子溶剂反应制得。另外,可以用含氟叔醇与碱反应,例如,氢氧化钾或氢化钠与含氟叔醇盐反应后可得到烷基化与烷基化剂含氟化合物。例如甲氧基纳米氟化丁烷或乙氧基纳米氟化丁烷。

Gupta 等详细分析和描述了含氟类发泡剂的制备方法。人们研究发现,这一类型的表面活性剂可以形成高泡沫质量及较高黏度的泡沫压裂液,其携砂性能良好,可以携带高质量的支撑剂(Gupta 等,2004)。

文献中的商品名称

文献	描述	出版商
Flow – Back™	黄胞胶,韦兰胶(Chatterji 等,2004)	Halliburton Energy Services, Inc.
Fluorinert™(Series)	氟碳化合物(Gupta 等,2004)	3M Comp.
Flutec™ PP	氟碳化合物(Gupta 等,2004)	BNFL Fluorochemicals Ltd.
Galden™ LS	氟碳化合物(Gupta 等,2004)	Montedison Inc.
Krytox™	氟化油和油脂(Gupta 等,2004)	DuPont
WS – 44	乳化剂(Welton 等,2010)	Halliburton Energy Services, Inc.

参 考 文 献

Bonekamp, J. E., Rose, G. D., Schmidt, D. L., Teot, A. S., Watkins, E. K., 1993. Viscoelastic surfactantbased foam fluids. US Patent 5 258 137, assigned to Dow Chemical Co., 2 November 1993.

Chatterji, J., Crook, R., King, K. L., 2004. Foamed fracturing fluids, additives and methods offracturing subterranean zones. US Patent 6 734 146, assigned to Halliburton Energy Services, Inc., Duncan, OK, 11 May 2004. < http://www.freepatentsonline.com/6734146.html >.

Chatterji, J., King, B. J., King, K. L., 2007. Recyclable foamed fracturing fluids and methods ofusing the same. US Patent 7 205 263, assigned to Halliburton energy Services, Inc., Duncan, OK, 17 April 2007. < http://www.freepatentsonline.com/7205263.html >. xmllabelb0020.

Dahanayake, M. S., Kesavan, S., Colaco, A., 2008. Method of recycling fracturing fluids using aself-degrading foaming composition. US Patent 7 404 442, assigned to Rhodia Inc., Cranbury, NJ, 29 July 2008. < http://www.freepatentsonline.com/7404442.html >.

Gupta, D. V. S., Pierce, R. G., SengerElsbernd, C. L., 2004. Foamed nitrogen in liquid CO_2 for fracturing. US Patent 6 729 409, 4 May 2004. < http://www.freepatentsonline.com/6729409.html >.

Harris, P. C., Heath, S. J., 1996. High-quality foam fracturing fluids. In: Proceedings Volume. SPEGas Technology Symposium, Calgary, Canada 28 April-1 May 1996, pp. 265-273.

Pakulski, M. K., Hlidek, B. T., 1992. Slurried polymer foam system and method for the use thereof. WO Patent 9 214 907, assigned to Western Co., North America, 3 September 1992.

Stacy, A. L., Weber, R. B., 1995. Method for reducing deleterious environmental impact ofsubterraneanfracturng processes. US Patent 5 424 285, assigned to Western Co., NorthAmerica, 13 June 1995.

Welton, T. D., Todd, B. L., McMechan, D., 2010. Methods for effecting controlled breakin pH dependent foamed fracturing fluid. US Patent 7 662 756, assigned to HalliburtonEnergy Services, Inc., Duncan, OK, 16 February 2010. < http://www.freepatentsonline.com/7662756.html >.

13 消 泡 剂

消泡对于一些工业来说非常有必要,通常是有效作业的关键因素。Owen 对消泡剂做了回顾(1996)。

13.1 消泡原理

13.1.1 泡沫的稳定性

泡沫是热力学不稳定体系,以下几个方面性能可防止泡沫塌陷:
(1)表面弹性;
(2)黏性疏水;
(3)减少气体在泡沫间的扩散;
(4)反相表面的相互作用带来的薄膜稳定效果。

泡沫的稳定性可以用吉布斯弹性形变 E 解释。在液膜扩展时,减少平衡中活性分子的表面浓度,就会产生吉布斯弹性形变,平衡表面张力 σ 随之增大,σ 也是恢复力。

$$E = 2A \frac{d\sigma}{dA} \tag{13.1}$$

式中,A 为表面积。在相互接触的泡沫中,具有时间依赖性的马仑高尼效应比较重要。由于在泡沫的扩大或收缩中表面活性剂的吸收速度有限,表面张力减小,就会出现与吉布斯弹性形变相关的恢复力。因此,马仑高尼效应是一种动力学效应。

膨胀模量可用来表示在非平衡条件下的表面张力效应。单一表面的复杂膨胀模量 ε 的定义方法与吉布斯弹性形变的定义方法相同。因子 2 不能应用于单一表面。

$$\varepsilon = 2A \frac{d\sigma}{dA} \tag{13.2}$$

在周期性膨胀实验中,复杂弹性模量是角频率的函数:

$$\varepsilon(i\omega) = |\varepsilon|\cos\theta + i|\varepsilon|\sin\theta = \varepsilon_d(\omega) + \omega\eta_d(\omega) \tag{13.3}$$

式中,ε_d 是膨胀弹性,η_d 是膨胀黏性。稳定泡沫的特点是高表面膨胀弹性和高膨胀黏性。因此,有效的消泡剂应能减小泡沫的这些性能。

在非平衡条件下,较高的体积黏度和表面黏度能够延迟泡沫液膜变薄和泡沫破裂之前的拉伸变形。还有一个关于形成有序结构的问题,在液膜表面形成的有序结构可稳定泡沫。表面液晶相也可增强泡沫稳定性。

如果减少气泡之间的气体扩散,阻碍泡沫大小的改变以及泡沫大小改变带来的机械应力变化,能够延迟泡沫塌陷。因此,单一液膜比相应的泡沫存在时间更持久。然而在实际应用中,这种作用影响很小。电的影响,如在双电层内的影响,能够形成相对表面,对极薄的液膜比较重要,如液膜小于 10nm。特别是它们与离子型表面活性剂的作用。

13.1.2 消泡剂的作用

在高体积黏度下,表面张力的降低与泡沫稳定机制无关,但是,对于其他泡沫稳定机制,则有必要改变其表面性质。消泡剂能够改变活化泡沫的表面性质。大多数消泡剂的表面张力在 20~30mN/m 之间,一些消泡剂的表面张力见表 13.1。

表 13.1 一些消泡剂的表面张力

材料	表面张力,(mN/m 20℃时)
聚氧丙烯(3kDa)	31.2
聚二甲基硅氧烷(3.9kDa)	20.2
矿物油	28.8
玉米油	33.4
花生油	35.5
磷酸三丁酯	25.1

对于某些消泡剂配方的低表面张力,已经提出两个相关的消泡机理:

(1)在液体中,消泡剂分散在细小的液滴中,分子可以从液滴进入泡沫的表面;这种扩散产生的张力最终导致泡沫液膜破裂;

(2)此外,一般认为消泡剂分子形成单分子层而不是扩散,这种单分子层比原来在液膜上的单分子层连贯性差,导致液膜的不稳定。

扩散系数:

扩散系数的定义:发泡介质的表面张力 σ_f 与消泡剂的表面张力 σ_d、两种材料的界面张力 σ_{df} 的差值。

$$S = \sigma_f - \sigma_d - \sigma_{df} \qquad (13.4)$$

显然,随着消泡剂表面张力的变小,扩散系数不断增大。这就是消泡的热力学趋势。

以上是大部分不溶的液体消泡剂研究。然而,实验表明,具有一定分散性的疏水固体也可大大提高消泡效果。消泡剂的效果与有机硅处理过的二氧化硅接触角之间存在很强的关联性。在疏水二氧化硅去湿过程中发生直接的机械冲击,能够导致泡沫塌陷。

13.2 消泡剂的分类

为满足不同需求,消泡剂中要包含大量组分。可从不同角度对消泡剂进行分类,如消泡剂应用、消泡剂物质形态以及消泡剂化学性质。一般来说,消泡剂包含多种液态和固态的活性成分以及一些辅助剂例如乳化剂、铺展剂、增稠剂、防腐剂、携砂油、增溶剂、溶剂和水。

13.2.1 活性成分

活性成分是控制实际发泡的组成成分,可能是液体也可能是固体。

13.2.2 液体组分

降低表面张力是消泡剂最重要的物理性质,所以按分子疏水性对消泡剂进行分类比较合理。相反,通过官能团对有机化合物进行分类的依据通常是极性和亲水性,例如,在基础有机化学中常见的醇、酸和盐。作为液相组分的四类消泡剂是:

(1)碳氢化合物类;
(2)聚醚类;
(3)硅氧烷类;
(4)氟碳化合物类。

13.2.3 固体颗粒的协同消泡作用

被分散的固体只有在合适的配方中才能起到消泡作用。有些液体消泡剂也只有在固体存在时才起作用。一般认为,体系中的固体颗粒被表面活性剂携带至界面处,从而引起泡沫的不稳定。

当共同使用疏水性固体颗粒和不溶于泡沫溶液的液体时,就会发生协同消泡作用(Frye 和 Berg,1989)。针对很多泡沫体系使用的单组分消泡剂已有大量研究,通过分析较差的消泡效果,已阐明了无论是单独固体还是单独液体的液膜破裂机制。

13.2.4 有机硅消泡剂

聚二甲基硅氧烷在非水性体系中活性很好,在水基体系中几乎没有抑制泡沫的作用。但是当它和疏水改性二氧化硅混合使用时,即可成为一种高效消泡剂。

几个因素促成有机硅消泡剂的双重性质。例如,水溶性有机硅可集中在油气界面稳定泡沫,而硅分散相液滴在泡沫气液界面的快速传播过程中加速聚结,通过表面迁移造成液膜变薄。

在不同的油中,硅氧烷的溶解度都很低。实际上,缓慢的溶解速率取决于油的实际黏度和分散相液滴的浓度。气泡上升速度可用于计算界面流动性,从而揭示临界气泡的大小和硅氧烷在低浓度下快速聚结的原因。

13.2.5 典型组成

消泡剂和消泡组分均是用于消除水基压裂液中的泡沫(Zychal,1986)。一种水力压裂用的典型消泡剂组分见表13.2。

表13.2 一种水力压裂液用消泡剂的组成

组成	含量(%)
C_6—C_{12} 的极性化合物的混合物	50~90
无水山梨醇单油酸酯	10~50
聚乙二醇(3.8kDa)	10

原酸酯的加入能够产生酸,可用来降解泡沫。合适的原酸酯和聚原酸酯有原乙酸三甲酯、原乙酸三乙酯和相应的原甲酸酯。顺便说一句,聚原酸酯在医学上有很重要的应用价值(Heller 等,2002)。一些简单的原酸酯如图13.1所示。

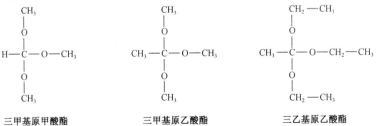

图13.1 原酸酯

由 Williamson 反应或者在氰化物中加入醇,都可以合成原酸酯。各自的反应如图 13.2 所示。

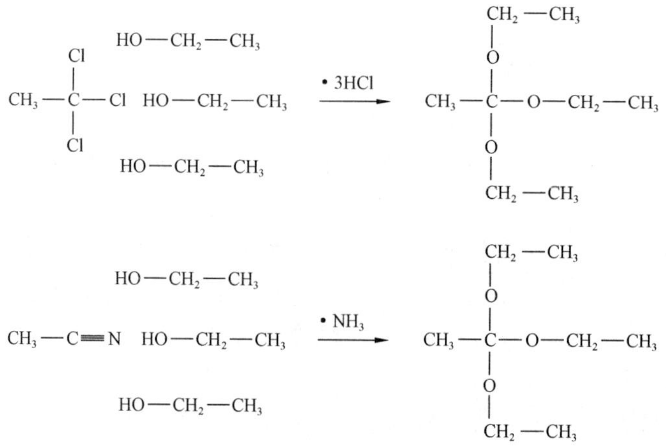

$$
CH_3-CCl_3 + 3\,HO-CH_2-CH_3 \longrightarrow CH_3-C(O-CH_2-CH_3)_3 + 3HCl
$$

$$
CH_3-C\equiv N + 3\,HO-CH_2-CH_3 \longrightarrow CH_3-C(O-CH_2-CH_3)_3 + NH_3
$$

图 13.2 原酸酯的合成

原酸酯对碱稳定,但是对酸和水不稳定。原酸酯能够降低泡沫压裂液的 pH 值,将发泡表面活性剂有效的转化为非发泡表面活性剂,从而使起泡的压裂液充分去除泡沫。在含水条件下原酸酯才能水解生成酸,可以是地层水也可以是向地层注入的水,应保证 1mol 原酸酯 2mol 水。

当原酸酯最终水解生成酸时,酸会与发泡表面活性剂发生反应生成非发泡的表面活性剂 (Welton 等,2010)。原酸酯组合物中可能含有抑制剂,该抑制剂可延迟原酸酯组合物中的原酸酯生成酸。此外,在延迟期间,抑制剂可能会中和生成的酸。抑制剂包括碱,例如碱金属氢氧化物、碳酸钠或六亚甲基四胺。

有时,相对于大量的弱碱,少量的强碱更利于延迟酸的生成,并且在要求的延迟时间内更利于中和生成的酸。易发泡的压裂液中还包含一些其他常用组分,例如(Welton 等,2010):

(1)稠化剂;
(2)杀菌剂;
(3)支撑剂。

文献中的商品名称

商品名	描述	供应商
WS-44	乳化剂(Welton et al,2010)	Halliburton Energy Services,Inc.

参 考 文 献

Frye,G. C. ,Berg,J. C. ,1989. Mechanisms for the synergistic antifoam action by hydrophobic solidparticles in insoluble liquids. J. Colloid Interface Sci. 130(1),54-59.

Heller,J. ,Barr,J. ,Ng,S. Y. ,Abdellauoi,K. S. ,Gurny,R. ,2002. Poly(ortho esters):synthesis,characterization,properties and uses. Adv. Drug Deliv. Rev. 54(7),1015-1039. <http://www. sciencedirect. com/science/article/

B6T3R - 46XH%20K44 - 4/2/fec170fd72f87dc13b7290e749af3388 >.

Mannheimer, R. J., 1992. Factors that influence the coalescence of bubbles in oils that containsilicone antifoamants. Chem. Eng. Commun. 113, 183 – 196.

Owen, M. J., 1996. Defoamers, fourth ed. In: Kirk – Othmer (Ed.), Encyclopedia of Chemical Technology, vol. 7. John Wiley and Sons, New York, Chichester, Brisbane, pp. 929 – 945.

Welton, T. D., Todd, B. L., McMechan, D., 2010. Methods for effecting controlled breakin pH dependent foamed fracturing fluid. US Patent 7 662 756, assigned to Halliburton Energy Services, Inc., Duncan, OK, 16 February 2010. <http://www.freepatentsonline.com/7662756.html>.

Zychal, C., 1986. Defoamer and antifoamer composition and method for defoaming aqueous fluidsystems. US Patent 4 631 145, assigned to Amoco Corp., 23 December 1986.

14 交 联 剂

14.1 交联反应动力学

羟丙基瓜尔胶与钛离子交联后，流变性更加复杂。为了更好地解释羟丙基瓜尔胶与钛螯合物交联后的流变性能，Barkat 研究了流变性随放置时间、剪切过程以及化学组成的变化（Barkat，1987）。

流变实验旨在获得羟丙基瓜尔胶的交联反应动力学信息。连续动态流动数据表明，对于交联剂和羟丙基瓜尔胶的浓度，交联反应级数约为 4/3 和 2/3。动态试验表明，冻胶的最终性能与剪切时间关系密切。

持续稳态剪切和动态实验表明，高剪切会对冻胶结构造成不可逆的破坏，随着剪切速率的增大，交联反应程度降低。剪切速率低于 $100s^{-1}$ 的实验表明，剪切过程会改变聚合物结构，从而影响化学反应和产物分子的性质。

为了降低流体的泵注摩阻，压裂液最好能够实现延迟交联，即降低交联反应速率。延迟交联的方法随后介绍。

14.2 交联剂

14.2.1 硼酸盐系列

硼酸能够与羟基化合物进行反应，如硼酸和甘油的反应机理如图 14.1 所示。三个羟基单元形成酯，一个单元形成复杂链接，同时产生一个质子，pH 值降低。该机理也适用于聚羟基化合物，通过同样的链连接两种聚合物。

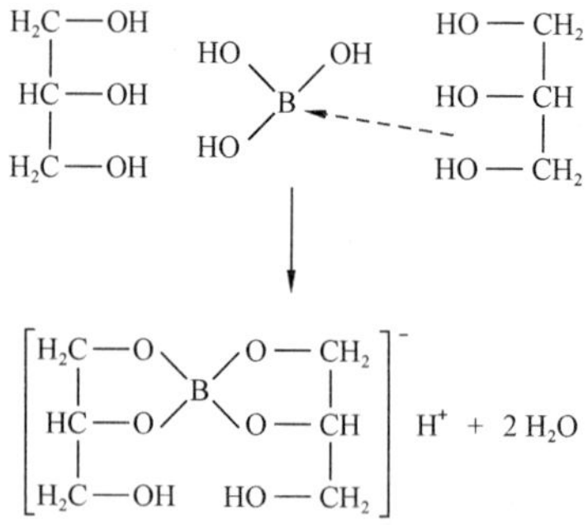

图 14.1 硼酸和甘油形成的络合物

控制延迟交联时间，则需要控制 pH 值、有效硼酸盐离子的释放或者同时控制两种因素。在淡水体系中可有效控制 pH 值（Ainley 等，1993）。使用微溶硼酸盐或者硼酸盐与多种有机物的络合物，可有效控制硼酸盐离子在淡水和海水中的释放。

硼交联压裂液已成功用于压裂作业。这种液体在高于 105℃ 时，有很好的流变性能、滤失性能,并产生良好的裂缝导流能力。硼交联是一个平衡过程,可在低剪切速率下具有较高的液体黏度(Cawiezel 和 Elbel,1990)。

硼交联压裂液由以下方式制备(Harris 等,1994):

制备 14.1:在(海)水中加入一种多糖聚合物形成凝胶,然后加入碱,调节 pH 值大于 9.5,最后加入硼酸盐交联剂形成交联冻胶。

干燥的颗粒物可通过以下方式制备(Harris 和 Heath,1994):

制备 14.2:在水溶液中加入 0.2%~1.0% 的水溶性多糖,形成凝胶,再加入硼化合物进行混合,形成硼交联多糖,最后进行干燥、造粒。

硼酸盐交联剂可以是硼酸、硼砂、碱土金属硼酸盐、碱金属碱土金属硼酸盐。硼的量以氧化硼计算,必须保证 5%~30%。

含硼的淀粉化合物可用于控制压裂用水合聚合物的交联速率。在水溶液中,淀粉和硼化合物反应可生成含硼淀粉化合物,该化合物可提供硼酸盐离子,能够交联水溶液中的水合聚合物(Sanner 等,1996)。在较低的温度条件下起到延迟交联的作用。

在相邻碳原子上含有顺式羟基或在 1,3 位上含有羟基的有机多羟基化合物可与硼酸盐反应生成五元或六元环络合物,随着 pH 值的变化,该反应是可逆的。

优选聚合物和硼酸盐阴离子的浓度,可获得我们所需要的冻胶。能够控制交联时间的可溶性硼酸盐溶液是非常好的交联剂。因为微溶的硼酸盐悬浮液可调节交联时间,因此也可用作水力压裂用交联剂(Dobson 等,2005)。硼酸盐矿物见表 14.1。

表 14.1　难溶硼酸盐矿物(Mondshine,1986;Dobson 等,2005;Parris 和 ElKholy,2009)

矿物名称	化学式
基性硼钠钙石	$NaCaB_5O_9 \cdot 5H_2O$
硼钠钙石	$NaCaB_5O_9 \cdot 8H_2O$
四水硼钙石	$CaB_6O_{10} \cdot 4H_2O$
戈硼钙石	$CaB_6O_{10} \cdot 5H_2O$
水硼钙石	$Ca_2B_4O_8 \cdot 7H_2O$
硬硼钙石	$Ca_2B_6O_{11} \cdot 5H_2O$
三斜硼钙石	$Ca_2B_6O_{11} \cdot 7H_2O$
板硼钙石	$Ca_2B_6O_{11} \cdot 13H_2O$
白硼钙石	$Ca_4B_{10}O_{19} \cdot 7H_2O$
纤硼钙石	$Ca_4B_{10}O_{19} \cdot 20H_2O$
基性硼钙石	$Ca_2B_{14}O_{23} \cdot 8H_2O$
柱硼镁石	$MgB_2O_4 \cdot 3H_2O$
水硼镁石	$MgB_2O_{13} \cdot 4H_2O$
库水硼镁石	$Mg_1B_6O_{11} \cdot 15H_2O$
多水硼镁石	$Mg_2B_6O_{11} \cdot 15H_2O$
斜方水硼镁石	$Mg_3B_{10}O_{18} \cdot 4H_2O$
水方硼石	$CaMgB_6O_{11} \cdot 6H_2O$
水硼镁钙石	$CaMgB_6O_{11} \cdot 11H_2O$
硼钾镁石	$KMg_2B_{11}O_{19} \cdot 9H_2O$
水硼锶石	$SrB_6O_{10} \cdot 2H_2O$

硼酸盐交联液流变性能的振荡实验测试表明(Edy,2010),线性黏弹性区和屈服频率与温度相关,并且屈服频率随温度升高呈指数上升。

屈服点定义为储能模量 G' 和耗能模量 G'' 相等时的角频率。在这一点,液体从弹性为主转化为以黏性为主。

14.2.2 钛化合物

有机钛化合物也可作为交联剂使用(Putzig 和 Smeltz,1986;Putzig,2010a)。水溶性钛组分通常是钛化合物的混合物,合适的有机钛化合物见14.2。

表 14.2 有机钛化合物(Putzig,2010a)

化合物类型	化合物类型
醇胺钛络合物	二乙醇胺乳酸钛
二乙醇胺钛络合物	三乙醇胺乳酸钛
三乙醇胺钛络合物乳酸钛	二异丙胺乳酸钛
乙二醇钛	乳酸钠钛盐
乙酰丙酮钛	山梨醇钛络合物
乳酸铵钛	

由于产量受限,油井服务公司缺少用于提高油气产量或采收率的,可与钛交联配合使用的延迟剂(Putzing,2010a)。

羟烷基氨基羧酸可作为延迟剂使用。优选的延迟剂为双(2-羟乙基)甘氨酸。该化合物如图14.2所示。

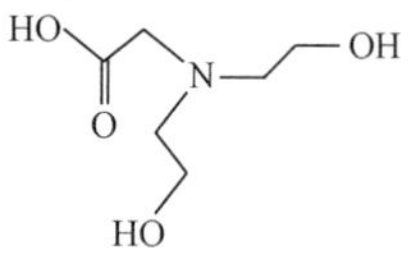

图 14.2　N,N-二(2-羟乙基)甘氨酸

14.2.3 锆化合物

用作延迟交联剂的锆化合物见表14.3。低分子化合物生成的络合物与分子间多糖化合物交换,起到延迟交联作用。

表 14.3 可用作延迟交联剂的锆化合物

锆交联剂/螯合剂	参考文献
羟乙基-三羟丙基乙二胺[①]	Putzig(1988)
卤化锆螯合物	Ridland 和 Brown(1990)
硼锆螯合物[②]	Dawson 和 Le(1998);Sharif(1995)

① 高温稳定性好。
② 适用于高温情况,增强压裂液稳定性。

二氨基化合物如图 14.3 所示;羧酸如图 14.4 所示;适合与锆化合物络合的多羟基化合物如图 14.5 所示。

图 14.3 羟乙基三羟丙基乙二胺

$HOCH_2COOH$

乙醇酸

乳酸

甘油酸

苯基乙醇酸

葡萄糖酸

柠檬酸

酒石酸

羟基丁酸

图 14.4 羟基酸

阿拉伯糖醇

己六醇

内消旋赤藓糖醇

甘油

内消旋环己六醇

内消旋甘露糖醇

山梨糖醇

图 14.5 用于络合的多元醇

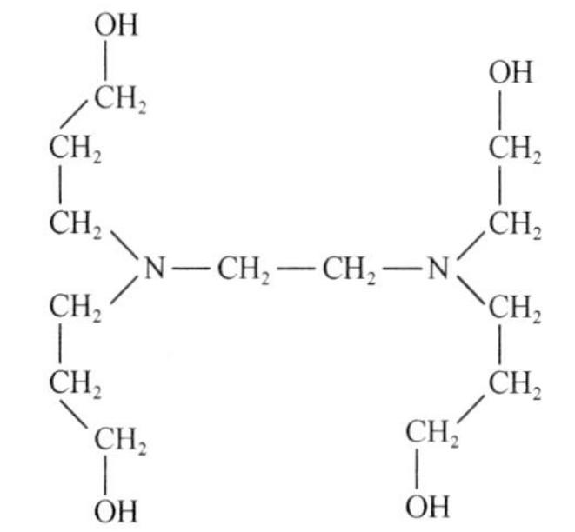

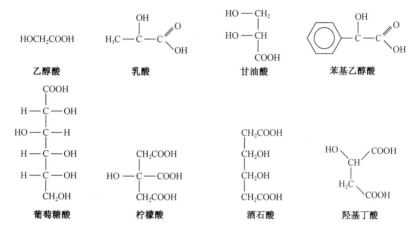

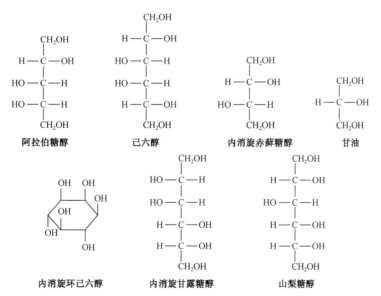

硼锆配合物可由 4-N-丙基锆与三乙醇胺和硼酸反应制得(Putzig,2010b)。硼锆络合物可在 pH 值为 8~11 范围内使用。

14.2.4 瓜尔胶交联

疏水改性瓜尔胶可用于钻井液、完井液或者修井液(Audibert 和 Argillier,1998)。改性瓜尔胶与聚合物或活性黏土一起使用。

将聚烷氧基烷基烯化酰胺接枝到瓜尔胶分子上得到水溶性瓜尔胶衍生物(Bahamdan 和 Daly,2007)。研究这些产品的流变性能,在高温高压下测量黏度目的是模拟井下条件。用乳酸锆交联后具有更好的高温稳定性和更高的冻胶黏度。

交联冻胶的目的是携带大量的支撑剂。为了促进冻胶从地层中返排出来,使用酶破胶剂对疏水改性瓜尔胶进行降解,用甲苯萃取冻胶碎片可生成稳定的乳液。用这种方式,通过对冻胶水解过程中产生的冻胶碎片进行乳化来促进液体返排。

在不增加聚合物用量的情况下,瓜尔胶和交联技术的改革促进了高黏硼交联压裂液的发展。对于低聚合物硼交联压裂液来说,之前很多地层都是温度过高或者过深,但现在已得到成功应用(Kostenuk 和 Gagnon,2008)。

从历史上看,对于加拿大西部沉积盆地地层,温度大于 80℃,地层深度超过 2500m,需要使用的聚合物浓度为 3.6~4.2kg/m³,而现在仅需要使用浓度为 1.8kg/m³ 的低聚合物浓度压裂液进行压裂,并具有明显的效果。

对压裂总体设计更改最少的情况下,使用低聚合物压裂液代替高聚合物压裂液。一种性能优良的压裂液由于其良好的剪切和温度稳定性,可以在传统泵速和支撑剂浓度条件下实现在线泵注。

低聚合物压裂液的优势包括:提高产量、减少成本及降低摩阻。夏天,低聚合物压裂液可用于温度高于 100℃,深达 3250m 的地层(Kostenuk 和 Gagnon,2008)。

羟丙基瓜尔胶。

羟丙基瓜尔胶凝胶可与硼酸盐(Miller 等,1996)、钛酸盐或锆酸盐交联。硼交联压裂液和线性羟丙基纤维素凝胶是最常用的高渗透储层用压裂液。这些凝胶用于高温高剪切条件下的水基压裂液中。

14.2.5 延迟交联剂

在一定 pH 值范围内,乙二醛(Dawson 1992a,1992b)是有效的延迟剂。乙二醛如图 14.6 所示,它与硼酸和硼酸盐离子发生化学结合,抑制最初用于水合多糖交联溶液中的硼酸盐离子的数量(例如半乳甘露聚糖)。

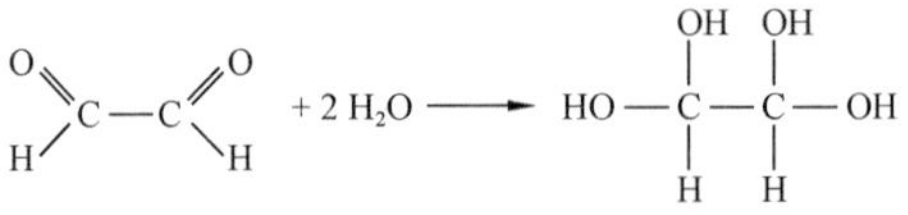

$$\underset{H}{\overset{O}{\underset{\|}{C}}}-\underset{H}{\overset{O}{\underset{\|}{C}}} + 2H_2O \longrightarrow HO-\underset{H}{\overset{OH}{\underset{|}{C}}}-\underset{H}{\overset{OH}{\underset{|}{C}}}-OH$$

图 14.6 乙二醛水合反应

多糖的交联速率可通过调节溶液的 pH 值控制。延迟交联机理如图 14.7 所示。如果两个低分子羟基化合物与高分子化合物交换,羟基属于不同的分子,就会发生交联作用。

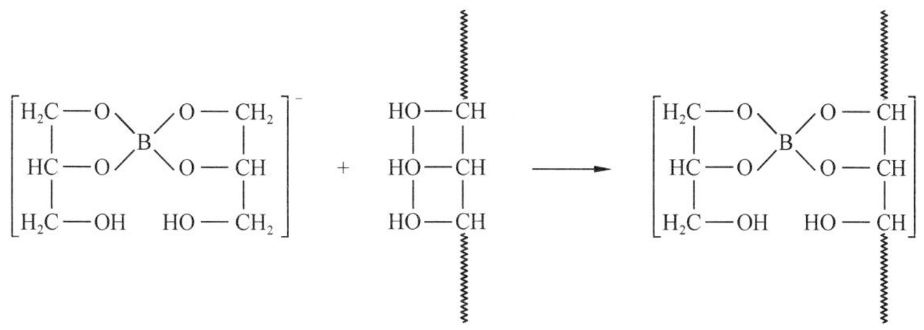

图 14.7 延迟交联

其他二醛、酮醛、羟基醛、邻位取代的芳香族二醛和邻位取代的芳香族羟基醛被证明以类似的方式进行反应(Dawson,1992a)。在高温条件下,淡水和海水均可使用硼交联瓜尔胶压裂液。

通过加入氟离子,产生不溶氟化镁沉淀,形成半乳甘露聚糖压裂液体系,该体系采用氧化镁延迟硼酸交联剂,增加使用温度范围(Nimerick 等,1993)。

此外,可能会加入镁离子螯合剂。使用氟化镁沉淀或者镁离子的螯合,在升高的温度下难生成不溶性氢氧化镁,pH 值降低,使硼酸盐交联逆向反应。这种添加剂可有效扩大该压裂液的使用温度范围至 135~150℃。

多元醇如乙二醇或甘油,能够延迟半乳甘露聚糖的水基压裂液硼酸交联反应(Ainley 和 McConnell,1993),可应用于高达 150℃ 的高温。最初形成低分子量的硼酸盐化合物,但是与凝胶的羟基交换缓慢。

参 考 文 献

Ainley, B. R., McConnell, S. B., 1993. Delayed borate crosslinked fracturing fluid. EP Patent 528 461, assigned to Pumptech NV and Dowell Schlumberger SA, 24 February 1993.

Ainley, B. R., Nimerick, K. H., Card, R. J., 1993. High – temperature, borate – crosslinked fracturingfluids: A comparison of delay methodology. In: Proceedings Volume. SPE Prod. Oper. Symp., Oklahoma, City, 21 – 23 March 1993, pp. 517 – 520.

Audibert, A., Argillier, J. F., 1998. Process and water – base fluid utilizing hydrophobically modifiedguars as filtrate reducers. US Patent 5 720 347, assigned to Inst. Francais Du Petrole, 24February 1998.

Bahamdan, A., Daly, W. H., 2007. Hydrophobic guargumderivatives prepared by controlled grafting processes – part II: rheological and degradation properties toward fracturing fluids applications. Polym. Adv. Technol. 18(8), 660 – 672. http://dx.doi.org/10.1002/pat.875.

Barkat, O., 1987. Rheology of flowing, reacting systems: the crosslinking reaction of hydroxypropyl guar with titanium chelates. Ph. D. Thesis, Tulsa University.

Cawiezel, K. E., Elbel, J. L., 1990. A new system for controlling the crosslinking rate of borate fracturing fluids. In: Proceedings Volume. 60th Annual SPE California Reg Mtg., Ventura, California, 4 – 6 April 1990, pp. 547 – 552.

Dawson, J. C., 1992a. Method and composition for delaying the gellation (gelation) of borated galactomannans. US Patent 5 082 579, assigned to BJ Services Co., 21 January 1992.

Dawson, J. C., 1992b. Method for delaying the gellation of borated galactomannans with a delayadditive such as glyoxal. US Patent 5 160 643, assigned to BJ Services Co., 3 November 1992.

Dawson, J. C., Le, H. V., 1998. Gelation additive for hydraulic fracturing fluids. US Patent 5 798 320, assigned to BJ

Services Co. ,25 August 1998.

Dobson Jr. ,J. W. , Hayden, S. L. , Hinojosa, B. E. ,2005. Borate crosslinker suspensions with moreconsistent crosslink times. US Patent 6 936 575, assigned to Texas United Chemical Co. , LLC. , Houston, TX,30 August 2005. < http://www. freepatentsonline. com/6936575. html > .

Edy, I. K. O. ,2010. Rheological characterization of borate crosslinked fluids using oscillatory measurements. University of Stavanger, Stavanger, Norway, Master's Thesis.

Harris, P. C. , Heath, S. J. ,1994. Delayed release borate crosslinking agent. US Patent5 372 732, assigned to Halliburton Co. , Duncan, OK,13 December 1994. < http://www. freepatentsonline. com/5372732. html > .

Harris, P. C. , Norman, L. R. , Hollenbeak, K. H. , 1994. Borate crosslinked fracturing fluids. EP Patent594 363, assigned to Halliburton Co. ,27 April 1994.

Kostenuk, N. , Gagnon, P. , 2008. Polymer reduction leads to increased success: a comparative study. SPE Drill. Completion 23(1) ,55 – 60. http://dx. doi. org/10. 2118/100467 – PA.

Miller, II. , W. K. , Roberts, G. A. , Carnell, S. J. , 1996. Fracturing fluid loss and treatment designunder high shear conditions in a partially depleted, moderate permeability gas reservoir. In: Proceedings Volume. SPE Asia Pacific Oil & Gas Conf. , Adelaide, Australia,28 – 31 October1996, pp. 451 – 460.

Mondshine, T. C. ,1986. Crosslinked fracturing fluids. US Patent 4 619 776, assigned to TexasUnited Chemical Corp. (Houston, TX) ,28 October 1986. < http://www. freepatentsonline. com/4619776. html > .

Nimerick, K. H. , Crown, C. W. , McConnell, S. B. , Ainley, B. , 1993. Method of using boratecrosslinked fracturing fluid having increased temperature range. US Patent 5 259455,9 November 1993.

Parris, M. D. , ElKholy, I. ,2009. Method and composition of preparing polymeric fracturing fluids. US Patent 7 497 263, assigned to Schlumberger Technology Corp. , Sugar Land, TX,3 March2009. < http://www. freepatentsonline. com/7497263. html > .

Putzig, D. E. ,1988. Zirconium chelates and their use for cross – linking. EP Patent 278 684, assignedto Du Pont De Nemours & Co. ,17 August 1988.

Putzig, D. E. ,2010a. Cross – linking composition and method of use. US Patent 7 732 382, assignedto E. I. du Pont de Nemours and Co. , Wilmington, DE,8 June 2010. < http://www. freepatentsonline. com/7732382. html > .

Putzig, D. E. ,2010b. Process to prepare borozirconate solution and use as cross – linker in hydraulicfracturing fluids. US Patent 7 683 011,23 March 2010. < http://www. freepatentsonline. com/7683011. html > .

Putzig, D. E. , Smeltz, K. C. ,1986. Organic titanium compositions useful as cross – linkers. EP Patent195 531,24 September 1986.

Ridland, J. , Brown, D. A. ,1990. Organo – metallic compounds. CA Patent 2 002 792,16 June 1990.

Sanner, T. , Kightlinger, A. P. , Davis, J. R. ,1996. Borate – starch compositions for use in oil fieldand other industrial applications. US Patent 5 559 082, assigned to Grain Processing Corp. ,24 September 1996.

Sharif, S. , 1995. Process for preparation of stable aqueous solutions of zirconium chelates. USPatent 5 466 846, assigned to Benchmark Res. &Technl In,14 November 1995.

15 冻胶稳定剂

为了防止交联的冻胶由于二价或三价金属离子的污染而降解,需要加入冻胶稳定剂,这对于高温压裂作业尤为重要。

15.1 化学物质

常用的冻胶稳定剂见表15.1。

表15.1 冻胶稳定剂

化合物	参考文献
硫代硫酸钠	Dawson 和 Le(1998),Willberg 和 Nagl(2004),Lord 等(2005)
葡萄糖酸钠	Crews(2006)
葡庚糖酸钠	Crews(2006)
二乙醇胺	Crews(2006)
三乙醇胺	Crews(2006)
甲醇	Crews(2006)
羟基甘氨酸乙酯	Crews(2006)
四乙烯戊胺	Crews(2006)
乙二胺	Crews(2006)

硫代硫酸钠是一种脱氧剂(Gupta,Carman,2010a)。脱氧剂通常是还原剂,可以将氧分子还原成 -2 价的低价态,并形成化合物,以此除去水中溶解的氧气。被还原的氧与离子、原子结合,形成含氧化合物。还原剂与氧气反应时会产生热量,必须在低温下进行反应。

已有实验证明,不宜使用高温硫代硫酸盐作为冻胶稳定剂。如果必须使用高温的稳定剂,通常使用三乙醇胺作为替代品。其他不含硫的高温冻胶稳定剂包括甲醇、乙醇胺、乙二胺、正丁胺以及这些化合物的混合物(Crews,2007)。

葡萄糖酸钠、葡庚糖酸钠常用于钙、镁、铁、锰和铜等阳离子的螯合作用,它们在复杂的钛酸盐、锆和硼酸盐离子的延迟交联方面的应用也被认可。此外,铁络合剂对破胶剂的稳定性和酶交联冻胶的稳定性方面的作用也非常明显(Crews,2006)。

15.2 特殊问题

15.2.1 水软化剂

冻胶稳定作用也可以在硬混合水中加入软化剂来实现。硬混合水指 $CaCO_3$ 总溶解量有

1000mg/L 以上的混合水。

通常油田水溶解的固体总含量为 3000～7000mg/L,水软化剂会把游离状态的高价金属离子浓度降低到 $CaCO_3$ 当量的 3000mg/L 以下。

优选水软化剂的作用类似于螯合剂,一般是无机磷酸盐的盐或者氨基羧酸,例如乙二胺四乙酸、聚丙烯酸、典型的磷酸盐阻垢剂。二乙烯三胺五钠盐是一种性能优异的水软化剂(Le 和 Wood,1993)。

15.2.2 硼酸盐的制备

已经在硼交联瓜尔胶体系中使用缓慢溶解的金属氧化物,以此增大液体中的碱度,促进交联反应,或者直接水中加入在低溶解能力的硼酸钙盐。此外,难溶硼酸盐的使用在一定程度上增强了冻胶的热稳定性,因为生成的硼离子可在一段时间内持续交联。

现在已经开发出控制交联和耐高温硼压裂液冻胶稳定剂的络合剂。液体和在硼酸盐离子存在时能够成胶的聚合物混合配制成基液,能在硼酸盐离子溶液中交联的交联剂和延迟添加剂混合制得络合剂。

在选定的 pH 值范围内,延迟添加剂有利于与硼和硼酸盐离子形成化学键,硼酸盐离子是由交联剂产生的,调节溶液中游离硼酸盐离子的浓度,以便于和多糖发生交联。另外,为了能更精确地控制延迟时间,络合剂中含有富裕的硼离子,有利于冻胶在高温状态下的稳定性。制备的详细过程如下。

制备 15.1:在水中加入乙二醛形成 300 份 40% 的溶液,边搅拌边加入 130 份十水合硼酸盐,形成乳白色悬浮液,渐变为 65 份 25% 的黄色溶液,溶液 pH 值范围是 4.9～6.5。然后加入 71.4 份 70% 的山梨醇溶液,加热到 95℃ 并保持 3h,加热过程中,溶液颜色从浅黄色到琥珀色。冷却到环境温度之后,溶液 pH 值范围是 4.5～5。1gal❶ 络合剂含有硼当量浓度为硼元素 0.29 lb,或者硼酸 1.65 lb。

络合剂也可以用于双重交联体系,混合传统硼交联剂和络合剂,可以得到更快的交联时间。所以在高温条件下,加入少量的硼或者硼酸钠,可以增强体系性能。当压裂液从井口通过油管注入地层输砂时,少量的传统交联剂可以增大压裂液黏度。增大的黏度不会影响压裂液的其他性能(Dawson,1992)。

15.2.3 电子供体化合物

加入冻胶稳定剂,减少冻胶热降解的化合物中,用肼作为供电子体已见诸报道(Pakulski Gupta,1994)。这种供电子化合物能够使冻胶在高达 150℃ 的条件下稳定。

研究发现,包含氧或者硫的杂环化合物有利于冻胶耐更高温度,一般应用在油田操作上。此类化合物能够在 205℃ 的条件下阻止冻胶的热降解。反应动力学在低温条件下也足够快,以使冻胶稳定。杂环化合物中氧原子的空间位阻很小,含有的独立电子对能够为冻胶稳定剂提供电子(Gupta 和 Carman,2010b)。

氧族杂环供电子化合物总结见表 15.2。

---

❶ 1gal(美) = $3.785412dm^3$。

表 15.2 杂环供电子化合物(Gupta 和 Carman,2010a)

化合物	化合物
四氢呋喃	甲基四氢呋喃
2-甲基呋喃	2-甲基-5-甲硫基呋喃
二糠基硫醚	二苯并呋喃
1,2,3,4-四氢二苯呋喃	2-酰基-5-甲基呋喃
溴化四氢二苯呋喃	2,3-二氢呋喃
2,2-二甲基四氢呋喃	2,5-二甲基四氢呋喃
2,3,4,5-四甲基呋喃	2-甲基-5-丙酰基呋喃
3-乙酰基-2,5-二甲基呋喃	2-乙酰基呋喃
2-乙酰基-2,5-二甲基呋喃	顺丁烯二酸酐
噻吩	丁二酸酐

氧族杂环供电子化合物可以和常规的除氧剂结合使用(Gupta 和 Carman,2010a)。常见的此类化合物见表 15.3。实际应用时,氧族杂环供电子化合物与常规除氧剂的质量比范围为 0.01~1,通常选 0.1~0.2。

氧族杂环供电子化合物可以提前加入水中,也可以和其他添加剂一起加入水基冻胶。如果其他操作条件允许,也可以加入到相邻设备中(Gupta 和 Carman,2010b)。

表 15.3 除氧剂

化合物	化合物	化合物
硫代硫酸钠	亚硫酸钠	亚硫酸氢钠
焦性没食子酸	联苯三酚	邻苯二酚
异抗坏血酸钠	维生素 C	间苯二酚
氯化亚锡	醌	对苯二醌

15.2.4 pH 值对冻胶稳定剂的影响

制备标准的油基冻胶体系,并加入不同用量的酸或碱。用 3.12mmol 的碱式氯化铝,然后在 300mL 柴油里加入 1.8mL LO-11A-LA,形成 50% 的溶液,配制成铝基体系。铁冻胶是在 300mL 柴油中加入 6.75mmol 硫酸铁和 1.8mL LO-11A-LA 制备。

酸的加入很明显对两种体系的流变和耐温性都不利,但这种铁基的加入却使流变性有很大提升。铁体系比酸或者碱体系的耐温性要更优异(Lawrence Warrender,2010)。

文献中的商品名称

名称	描述	供应商
Alkaquat™ DMB-451	烷基二甲基苄基氯化铵(Lawrence 和 Warrender,2010)	Rhodia Canada Inc
Geltone^R (Series)	有机黏土(Lawrence 和 Warrender,2010)	Halliburton Energy Services, Inc.
Rhodafac^R LO-11A-LA	磷酸酯(Lawrence 和 Warrender,2010)	Rhodia Inc. Corp.

参 考 文 献

Crews, J. B. , 2006. Biodegradable chelant compositions for fracturing fluid. US Patent 7 078 370, assigned to Baker Hughes Inc. , Houston, TX, 18 July 2006. < http://www.freepatentsonline.com/7078370.html >.

Crews, J. B. , 2007. Fracturing fluids for delayed flow back operations. US Patent 7 256 160, assigned to Baker Hughes Inc. , Houston, TX, 14 August 2007. < http://www.freepatentsonline.com/7256160.html >.

Dawson, J. C. , 1992. Method for improving the high temperature gel stability of borated galactomannans. US Patent 5 145 590, assigned to BJ Services Co. , Houston, TX, 8 September 1992. < http://www.freepatentsonline.com/5145590.html >.

Dawson, J. C. , Le, H. V. , 1998. Gelation additive for hydraulic fracturing fluids. US Patent 5 798320, assigned to BJ Services Co. , 25 August 1998.

Gupta, D. V. S. , Carman, P. S. , 2010a. Method of treating a well with a gel stabilizer. US Patent 7767 630, assigned to BJ Services Co. , LLC, Houston, TX, 3 August 2010. < http://www.freepatentsonline.com/7767630.html >.

Gupta, D. V. S. , Carman, P. S. , 2010b. Method of treating a well with a gel stabilizer. USPatent Application 20100016182, 21 January 2010. < http://www.freepatentsonline.com/20100016182.html >.

Lawrence, S. , Warrender, N. , 2010. Crosslinking composition for fracturing fluids. US Patent 7 749946, assigned to Sanjel Corp. , Calgary, Alberta, CA, 6 July 2010. < http://www.freepatentsonline.com/7749946.html >.

Le, H. V. , Wood, W. R. , 1993. Method for increasing the stability of water-based fracturing fluids. US Patent 5 226 481, assigned to BJ Services Co. , Houston, TX, 13 July 1993. < http://www.freepatentsonline.com/5226481.html >.

Lord, P. D. , Terracina, J. , Slabaugh, B. , 2005. High temperature seawater-based cross-linked fracturing fluids and methods. US Patent 6 911 419, assigned to Halliburton Energy Services, Inc. , Duncan, OK, 28 June 2005. < http://www.freepatentsonline.com/6911419.html >.

Mondshine, T. C. , 1986. Crosslinked fracturing fluids. US Patent 4 619 776, assigned to Texas United Chemical Corp. , Houston, TX, 28 October 1986. < http://www.freepatentsonline.com/4619776.html >.

Pakulski, M. K. , Gupta, D. V. S. , 1994. High temperature gel stabilizer for fracturing fluids. US Patent 5 362 408, assigned to The Western Company of North America, Houston, TX, 8 November 1994. < http://www.freepatentsonline.com/5362408.html >.

Willberg, D. , Nagl, M. , 2004. Method for preparing improved high temperature fracturing fluids. USPatent 6 820 694, assigned to Schlumberger Technology Corp. , Sugar Land, TX, 23 November 2004. < http://www.freepatentsonline.com/6820694.html >.

16 破 胶 剂

压裂液把支撑剂携带到已形成的裂缝之后,一般需要通过使用破胶剂使其再生。破胶剂通常降低液体黏度到一定水平,使支撑剂停留在裂缝中,增大油井的裂缝接触程度。破胶剂通过减小其相对分子质量降解聚合物。裂缝成为高渗透的通道,油和气通过此路径进入井中(Armstrong,2012)。

破胶剂不仅能使冻胶破胶以达到再生的目的,更重要的是,破胶剂可以用于控制破胶的时间。过早破胶会使悬浮的支撑剂不能被液体带到裂缝中足够深的地方,过早破胶也会使流体黏度过早降低,导致不能压裂形成裂缝。

最好的方式是在泵送工作结束后,压裂冻胶开始破胶。为了更具实用性,冻胶应该在压裂过程结束后的特定时间例如24h内完全破胶。破胶之后的溶液可以通过洗井或者返排处理,从裂缝中回收(Armstrong,2012)。

16.1 水基体系的破胶

通常有两种混合压裂液和破胶剂的方法(Carpenter,2009):
(1)在压裂液泵入地下之前混合破胶剂和压裂液;
(2)先向地层中泵入压裂液,然后泵入破胶剂。

最起码第一种方法很方便,因为在地面上混合破胶剂和压裂液,以及向地下泵入混合物的操作都很简单,但这种方法的缺点就是在达到需要的时间之前,破胶剂就会降低压裂液的黏度。

第二种方法,先向地层中泵入压裂液,然后泵入破胶剂,在泵入破胶剂的时候很不方便,用这种方法,冻胶的黏度不会过早地降低。

压裂工作结束后,应该恢复裂缝的性能以最大化地提高油井的生产能力。能达到这种效果的唯一方法就是大幅度降低压裂液黏度和稠化剂相对分子质量,从而使压裂液降解。

羟丙基瓜尔胶常用的破胶剂有酶、氧化剂和催化氧化破胶剂等,已经有人发表了相关的降解动力学的综述。黏度的改变用时间来衡量。

研究表明,只有在不高于60℃的酸性介质中,酶破胶剂才能有效发挥作用。在温度不高于50℃的碱性介质中,催化氧化破胶体系最有效。50℃以上时,羟丙基瓜尔胶压裂液可以在没有催化剂的条件下用氧化破胶剂直接降解破胶。

16.2 氧化破胶剂

文献(Bielewicz和Kraj,1998)中描述了氧化破胶剂中的强碱、次氯酸盐以及无机和有机的过氧化物。这些物质通过氧化机理降解聚合物分子链。用羧甲基纤维素、瓜尔胶、部分水解聚丙烯酰胺等进行了一系列冻胶破胶的实验室研究。

16.2.1 次氯酸盐

次氯酸盐的氧化性较强,可以降解聚合物分子链,通常和叔胺一起使用(Williams 等,1987)。比单独使用时的反应速率快。叔胺半乳糖可以作为叔胺的来源。

破胶之前,也可以用作稠化剂。次氯酸盐也可以对稳定的流体进行破胶。硫代硫酸钠已经作为高温条件下的稳定剂应用。

16.2.2 过氧化物破胶剂

在含有羟丙基瓜尔胶的碱性溶液中,碱土金属的过氧化物常用作延迟破胶剂。升温之后,这些过氧化物就被激活。

研究表明,在低温条件下,过氧化钙比过氧化镁的破胶效果更好。过氧化镁更适用于高温条件下聚多糖液体的延迟破胶,过氧化钙则在相对较低的温度下效果更好。过磷酸盐的酯类和酰胺化合物可以用作氧化破胶剂,过磷酸盐会干扰交联反应的进行,而过磷酸盐的酯类和酰胺类化合物不会有影响。

用在深井施工的压裂液通常含有破胶剂,施工温度一般为 90~120℃,此时使用的交联剂为金属离子交联剂,例如钛和锆。基于压力的破胶剂也已经被报道(Harms,1992)。此外,有机过氧化物也适用于破胶过程。过氧化物不必完全溶于水。通过调节压裂液中破胶剂的量,可以控制破胶时间在 4~24h 内。

16.2.3 氧化还原破胶剂

一般来说,破胶剂的作用原理是氧化还原反应,二价铜离子和叔胺可以降解很多种聚多糖(Shuchart 等,1999)。

16.3 延迟释放的酸

在高渗透岩心中的羟乙基纤维素实验表明,过硫酸盐氧化破胶剂和酶破胶剂不能对此聚合物进行充分降解。过硫酸钠破胶剂降解聚合物的原理是热分解作用,矿物质的存在可以加速该过程。

酶破胶剂吸附在裂缝表面,仍有破胶剂的作用。在低 pH 值下用硼酸盐交联冻胶的动态滤失试验表明,可以用酸的延迟释放来控制钻井中的加速漏失。流变性能测试也表明,可溶性的酸延迟释放剂可以将硼酸盐交联液体转化成线性凝胶液(Noran,1995)。

乙醇酸固体:

当乙醇酸水凝胶用于压裂液时,乙醇酸固体可以用来减少液体用量(Cantu,Boyd,1989)。乙醇酸固体在裂缝形成的过程中降解,释放出乙醇酸,进行破胶。这个机理可以用来延迟破胶,如图 16.1 所示。裂缝的渗透性可以在不用把破胶剂分离的情况下恢复,固体产物作为降滤失剂。

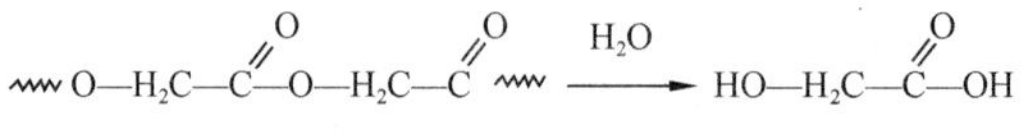

图 16.1 聚乙醇酸的水解

16.4 酶破胶剂

酶是催化剂和特殊的基质,会催化聚合物上的特定分子水解。用酶控制破胶可以避免氧化温度过高的问题,因为酶在低温下效果就很明显。在活性存在的条件下,酶可以降解很多聚合物链。但酶的有效pH值范围较窄,pH值太高可能会使酶失活。

在pH值为5~7的弱酸性条件下,传统酶降解半乳糖的作用效果最明显。现在已经研究出适用于特定聚合物的极端温度的酶(Brannon Tjon – Joe – Pin,1994;Sarwar等,2011;Jihua Sui,2011)。

关于酶活性的基础研究已经展开,研究者已经研究了降解产物、降解动力学和降解的应用,例如pH值、温度等(Slodki和Cadmus,1991)。因为酶降解过程具有高度的选择性,不存在通用的酶,但是对于某一种稠化剂,选择的酶在应用过程中必须有相当的成功率。适用于颗粒体系的酶列于表16.1。

表16.1 聚合物酶体系

聚合物	参考文献
黄胞胶[①]	Ahlgren(1993)
甘露聚糖纤维素[②]	Fodge等(1996)

① 高温和高盐浓度。
② 高碱度和温度。

酶适用于直接破坏稠化剂的分子链,也可以降解其他体系聚合物,这些聚合物通常降解为有机酸小分子,这些酸小分子会促进稠化剂的降解(Harris和Hodgson,1998a,1998b)。

尽管酶的性能优于传统氧化破胶剂,但酶破胶剂也有一定的局限性,因为酶会有相互影响,以及和其他添加剂不相容的问题。酶破胶剂和压裂液添加剂,例如杀菌剂、黏土稳定剂以及支撑剂表面树脂的相互作用已经有报道(Prasek,1996)。

16.5 胶囊破胶剂

胶囊破胶剂是用无法渗透或者轻微渗透的囊壁材料包裹破胶剂制备的,因此破胶剂最初不能和聚合物直接接触发生降解,只有当破胶剂从胶囊中释放出来或者胶囊破坏之后,破胶剂才会发生作用。

胶囊破胶剂广泛应用于延迟破胶。把破胶剂包装在一个抗水溶的胶囊内,这个胶囊把破胶剂和液体隔开,所以可以在压裂液中加入高浓度破胶剂,而不引起液体性能例如黏度等的过早损失。

设计胶囊破胶剂的关键因素是包裹材料的性能、释放机理和化学反应,例如可以用水降解的聚合物作为设计的包裹材料(Muir和Irwin,1999)。

这种延迟破胶的方法在氧化破胶和酶破胶方面都已经有报道。胶囊破胶剂的配方列于表16.2,可以利用的囊壁材料列于表16.3。

表 16.2　胶囊在延迟破胶中的应用

破胶体系	参考文献
过硫酸铵①	Gulbis 等（1990a,1990b,1992）；King 等（1990）
酶破胶剂②	Gupta 和 Prasek（1995）
混合组分③	Boles 等（1996）

① 瓜尔胶和纤维素衍生物。
② 开放的包装。
③ 用于钛和锆，木脂包装。

表 16.3　胶囊破胶剂的囊壁材料

囊壁材料	参考文献
聚酰胺①	Gupta 和 Cooney（1992）
交联的弹性纤维	Manalastas 等（1992）
和氮丙环预聚物或者碳二亚胺交联的部分水解丙烯酸树脂②	Hunt 等（1997）；NormanandLaramay（1994）；Norman 等（2001）
7% 沥青 +93% 中立化的磺化离聚物	Swarup 等（1996）

① 用于粒径 50~240μm 的过氧化物。
② 纤维素衍生物外面的涂层。

16.6　用于瓜尔胶的破胶剂

油气工业中，油气井水力压裂常用瓜尔胶冻胶体系作为黏性流体（Johnson,2011）。为了获得较高的渗透系数，压裂之后，必须要降解冻胶和滤饼。酶在此方面的应用很广泛，但高浓度的酶会导致过早降解，反而使冻胶失效，并且 pH 值和温度的苛刻要求限制了酶的应用。

只有在冻胶黏度和稠化剂充分降解后，井的产能才能达到最大化。压裂液黏度的降低和评价这些材料的传统方法不一定能表明冻胶组分已经彻底破胶。

已经研究了在 KCl 溶液中，羟丙基瓜尔胶和氧化剂（过硫酸铵）之间的反应，在特定的条件下，研究了溶液黏度和羟丙基瓜尔胶相对分子质量的变化。

当溴作为破胶剂时，用氨基磺酸稳定该破胶剂的活性是非常有效的，尤其是在 pH 值为 13 的时候。例如，WELLGUARD™7137 瓜尔胶破胶剂，如果避光保存，可以稳定一年。卤素化合物如溴化物、氯化物或者两者混合物可以作为卤素来源。

和次溴酸盐不一样，这些破胶剂不会氧化或者破坏常作为腐蚀剂阻垢剂的有机磷酸盐。此外，这种破胶剂对金属尤其是铁合金的腐蚀性较小。这是因为这种破胶剂氧化还原活性较低（Carpenter,2007,2009）。在 50℃ 的条件下研究了这种破胶剂对瓜尔胶的作用，如图 16.2 所示。

硼交联的瓜尔胶冻胶可以用 EDTA 破胶（Crews,2007a），EDTA 和其他氨基羧酸化合物可以使压裂液冻胶破胶。例子见表 16.4。

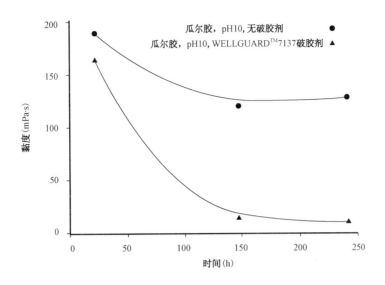

图 16.2　卤素破胶剂对瓜尔胶的影响

表 16.4　EDTA 相关的破胶剂（Crews,2007a）

络合物	络合物
四丙二胺四乙酸钠	乙二胺二乙酸二钠
三羟乙基乙二胺四乙酸钠	乙二胺二乙酸二钠钙二水合物
三氨基三乙酸钠	乙二胺四乙酸铵
三乙二胺三乙酸钠	

可以确信,这些破胶剂直接作用于聚合物本身,而不是任何的交联剂。同时,多烃化合物可以对瓜尔胶以及多糖形成的凝胶进行破胶。包括甘露醇和山梨醇。这些多元醇可以用在酶破胶剂的组合中(Crews,2007b)。

瓜尔胶的酶破胶剂:

在压裂施工中,瓜尔胶压裂液通常在 pH 值为 9.5～11 范围内交联,所以破胶剂需要能在该 pH 值范围内破胶,因此开发了水解酶(Armstrong,2012)。

水解酶的第 8 子族酶可能用于此处,这些酶包括来源于嗜碱菌 N16－5 的酶破胶剂,这些酶在特定的 pH 值条件下有最大的活性,这种 β －聚甘露糖酶的制备已经有文献描述(Ma,2004)。

这种酶从嗜碱菌 N16－5 的培养液中制备,活性最佳的条件是 pH 值 9.5,温度 70℃。由一个多肽的分子链组成,相对分子质量约为 55kDa,在半乳糖、葡甘露聚糖水解成低聚糖、和单糖的过程中,催化效果明显。

为了能用于压裂施工,要求在 15～107℃的温度范围内,这种酶破胶剂催化活性的耐温性要好(Tjon－Joe－Pin,1993)。

因为酶破胶剂在碱性 pH 值范围内活性有最大值,为了能在更大的 pH 值范围内更好地控制压裂液的水解作用,可以与其他在用于不同 pH 值范围的破胶剂联合使用。在交联的聚合物冻胶中,可以使用在 pH 值 4~8 的范围内耐温性较好的破胶剂,研究者已经总结描述了其中可以使用的酶(Tjon – Joe – Pin,1993)。

二价阳离子有可能促进或者抑制酶破胶剂的活性,Mg^{2+} 促进酶的活性,但 Co^{2+} 抑制酶的活性(Tjon – Joe – Pin,1993)。

可以在不同的聚酰亚胺(PEI)和硫酸葡聚糖(Cordova,2008)的比值下生产纳米颗粒,自行组装的聚酰亚胺和硫酸葡聚糖可以产生 100~200nm 粒径的颗粒,能够有效捕捉酶,并保持其在水中或者胶液中的胶体稳定性。加入氯化铬可以在几分钟之内形成冻胶,然而,铬离子在聚合电解质中处于游离状态,这也将显著延迟冻胶的形成。硼交联的瓜尔胶化学结构如图 16.3。

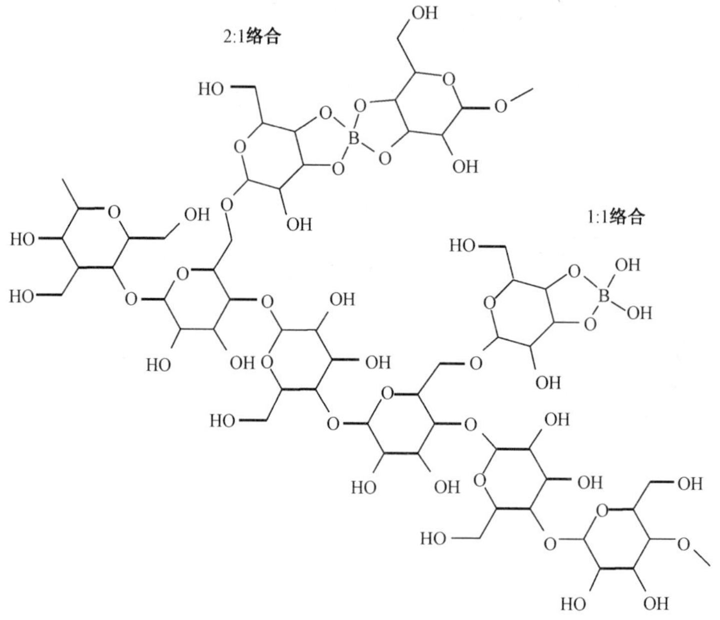

图 16.3 硼交联的瓜尔胶结构图

醚键很容易被酶,如果胶酶破坏(Barati,2011)。另一方面,聚酰亚胺和硫酸葡聚糖的聚合电解质复合体可以用于捕捉酶,达到控制释放,保护酶不受苛刻条件破坏的目的。

当酶保存在聚合电解质纳米颗粒中时,硼交联瓜尔胶冻胶流变性能的降解会推迟 11h,当酶不在该纳米颗粒中时,延迟时间是 3h。此外,聚酰亚胺和硫酸葡聚糖的聚合电解质可以保护酶不受高温和 pH 值的影响而失活(Johnson,2011;Barati,2012)。

氧化或者酶破胶剂已经用于研究时间、温度和破胶剂浓度与降解速率的关系(Kyaw,2012)。

化学破胶剂减小瓜尔胶聚合物黏度的机理,是把聚合物分解成小片段。黏度的降低有利于聚合物的返排,也就实现了聚合物从支撑剂填充层的快速回收。无效破胶剂或者破胶剂使用错误,可能导致脱砂或者随高黏流体的返吐,这两种结果都会大大降低油气井的产量。研究者也评估了破胶剂在低介质温度下的活性。

除了交联实验和破胶实验之外,其他几种方法也已经有详细的介绍(Kyaw,2012)。

可以用破胶实验之后的残余量确定破胶过程结束后未破胶的冻胶量,这是一种主要的滤失,可以测得不含破胶的聚合物的百分含量(Sarwar,2011)。

16.7 黏弹性表面活性剂凝胶液体

用黏弹性表面活性剂(VES)增稠的黏性流体,可以用脂肪酸盐的量控制破胶,例如硼交联的流体,用氧化胺表面活性剂增稠后,可以用含有从菜花籽油和玉米油中提炼的脂肪酸盐的组分破胶(Crews,2010)。

脂肪酸的转化或者皂化反应,可能发生在混合或者向地下泵入压裂液的过程中,此种方法也可以用在作业完成后最常发生皂化反应的油藏。或者预先制备好,后面作为后续破胶剂加入,移除已经被置入地下的 VES 凝胶液。

最可能发生的是压裂液黏度先上升,然后再降低。当菜籽油和 $Ca(OH)_2$ 混合,最开始 VES 流体的黏度会增加,然后是破胶反应(Crews,2010)。初始流体黏度的增加,可以解释为脂肪酸的部分皂化生成的表面活性剂使 VES 增稠。

16.8 颗粒剂

颗粒剂在延迟破胶方面可能有用。含有 40%～90% 的钠或者过硫酸铵破胶剂和 10%～60% 的无机黏合剂粉末,例如黏土的颗粒剂已经有报道(McDougall,1993)。颗粒剂可以延迟释放破胶剂。

其他作为延迟破胶的化学品也被称为可控水溶性化合物或者减缓某种盐释放的清除剂,聚磷酸盐就是其中一类(Mitchell,2001)。

颗粒剂是将固体破胶剂分散在蜡状基质中制成的,这种破胶剂在压裂中,用于磷酸烷基酯盐制备的烃液凝胶的破胶。蜡状颗粒剂在地表温度是固态,而地层温度下在液态烃中溶解或者扩散,释放破胶剂与稠化剂进行反应(Acker,Malekahmadi,2001)。

16.8.1 油基体系的破胶剂

无水体系用的破胶剂和有水基体系用的破胶剂化学性质完全不同。熟石灰和碳酸氢钠的混合物在无水体系的破胶过程中作用很大(Syrinek,Lyon,1989)。碳酸氢钠独自使用时,对磷酸铝基压裂液或者磷酸铝基凝胶剂的破胶完全无作用。一种替代方法就是用乙酸钠作为无水凝胶的破胶剂。

16.8.1.1 酶基破胶剂

用 VES 增黏的液体黏度可以在细菌、真菌或者酶等生物化学成分直接或间接的参与下减小。这些生物化学成分可以直接作用于 VES 本身,或者液体中其他副产物引起黏度的减小,也可能分解或者破坏 VES 凝胶液体的胶束结构,也可能根据某种机理产生一种酶,降低液体黏度。

一种单一的生物化学成分可能同时按两种机理反应,例如降解 VES,同时又有一部分,例如乙二醇后续产生副产物酒精引起黏度的下降。

或者同时可以使用两种或多种不同的生物化学成分。在特定的情况下,浓盐水用氧化胺表面活性剂凝胶化以后,可以在某些细菌存在的时候使其黏度下降,例如下水道的大肠杆菌、荧光假单胞菌和绿脓杆菌(Crews,2006)。

16.8.1.2 VES 的破胶增强剂

可以用油溶性表面活性剂作为 VES 凝胶液体的内部破胶剂的破胶增强剂(Crews 和 Huang,2010b)。油溶性表面活性剂的破胶增强剂可以克服盐度对内部破胶剂的减速影响,尤其是在低温的时候。此外,油溶性表面活性剂的破胶增强剂可以使 VES 凝胶在低浓度内部破胶剂条件下,快速彻底破胶。

油溶性表面活性剂的破胶稳定剂包括多种山梨醇酐(不饱和)、脂肪酸酯(Crews 和 Huang,2010a,2010b)。这些酯和矿油混合。研究发现,不饱和脂肪酸会自氧化降解成 VES 破胶产品或组分。和多烯酸混合后都显示出对 VES 表面活性剂胶束结构独特的降解,这些降解都是其自氧化反应产生的副产物导致的。

在这些自氧化过程中,会形成各种各样的氢过氧化物,这些反应的最终产物通常有羰基化合物、乙醇、酸和烃类。各种脂肪酸的自氧化速率列于表 16.5。

表 16.5 C_{18} 酸的自氧化相对速率

脂肪酸	双键数	自氧化相对速率
硬脂酸	0	1
油酸	1	100
亚油酸	2	1200
亚麻酸	3	2500

16.8.1.3 表面活性剂聚合物组分

将黏弹性表面活性剂单体和低聚物或者聚合的化合物形成的稳定骨架组合到一起,可以提高表面活性剂液体的黏弹稳定性。在这个骨架结构中,可以更换黏弹性表面活性剂的官能团(Horton,2009)。

黏弹性表面活性剂溶液可以用氯化 N-十二烯基-N,N-二(2-羟乙基)-N 甲基铵在液态氯化铵中制得。将氮气鼓入溶液以排空氧气,然后加入引发剂 2,2-偶氮(双脒基丙烷)二盐酸盐。图 16.4 简单介绍了该齐聚反应。同样的方式也可以制得十八烯酸钾。

$$CH_2 = CH-(CH_2)_{15}-COO^-K^+$$

这种低聚物产物可以作为低聚乙烯的骨架,在这个骨架上可以连接长的表面活性剂链。如果双键是共轭的,例如 1,3-十八二烯酸钾。产品低聚物的骨架结构和聚丁二烯相当。

$$CH_2 = CH-CH = CH-(CH_2)_{13}COOK^+$$

表面活性剂单体在胶束的影响下发生齐聚,使得凝胶对烃类化合物相当不敏感,并且,表面活性剂单体的齐聚反应很难影响表面活性剂凝胶的黏度。

其他的一些低聚的表面活性剂以及其共聚单体都被详细报道了(Horton,2007,2009)。图16.5是一个共同齐聚的例子。

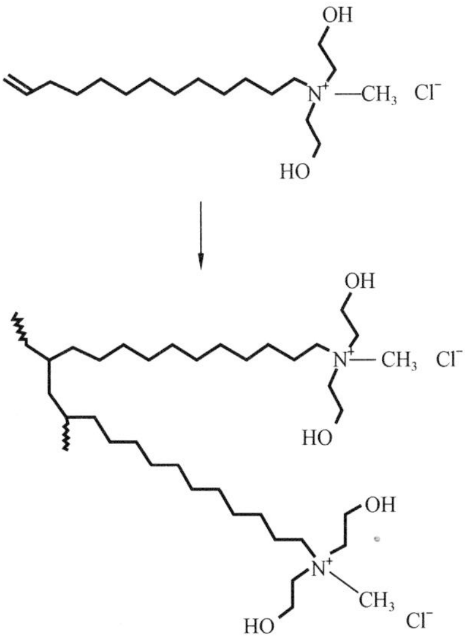

图 16.4　不饱和的四元化合物的齐聚

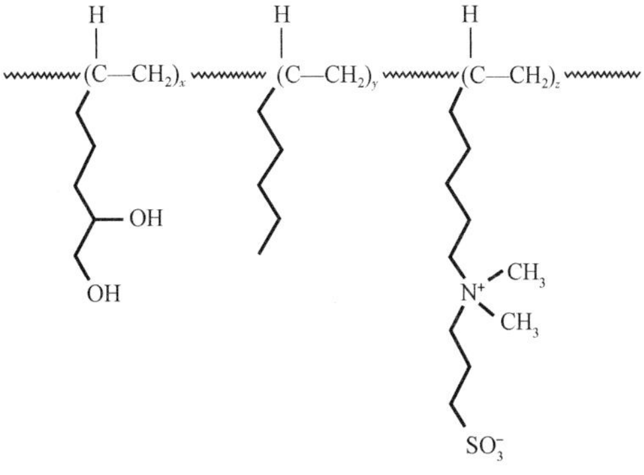

图 16.5　表面活性剂的共齐聚

邻位的二元醇很容易和高价的金属离子发生交联,关于液体流失的公式,可以用 API 标准加以描述(API StandardRP 13B – 1,2009)。

文献中的商品名称

商品名	描述	供应商
Benol®	石蜡油(Crews 和 Huang,2010a,2010b)	Sonneborn Refined Products
Captivates® liquid	鱼明胶和阿拉伯树胶胶囊(Crews 和 Huang,2010a,2010b)	ISP Hallcrest
Carnation®	石蜡油(Crews 和 Huang,2010a,2010b)	Sonneborn Refined Products
ClearFRAC™	增产措施工作液(Crews,2006,2010;Crews 和 Huang,2010a,2010b)	Schlumberger Technology Corp.
Diamond FRAQ™	VES 破胶剂(Crews 和 Huang,2010b)	Baker Oil Tools
DiamondFRAQ™	VES 体系(Crews,2010;Crews 和 Huang,2010a,2010b)	Baker Oil Tools
Escaid®(Series)	矿物油(Crews 和 Huang,2010a,2010b)	Crompton Corp.
Hydrobrite® 200	石蜡油(Crews 和 Huang,2010a,2010b)	Sonneborn Inc.
Isopar®(Series)	异烷烃溶剂(Crews 和 Huang,2010a,2010b)	Exxon
Jordapon® ACI	椰油基羟丙基磺酸钠表面活性剂(Crews,2010)	BASF
Jordapon® CI	椰油酰羟乙磺酸酯铵表面活性剂(Crews,2010)	BASF
Microsponge™	多孔物质(Crews,2006,2010;Crews 和 Huang,2010a,2010b)	Advanced Polymer Systems
Poly – S. RTM	聚合物胶囊(Crews,2006,2010;Crews 和 Huang,2010a,2010b)	Scotts Comp.
Span® 20	失水山梨醇月桂酸酯(Crews 和 Huang,2010b)	Uniqema
Span® 40	失水山梨醇单棕榈酸酯(Crews 和 Huang,2010b)	Uniqema

续表

商品名	描述	供应商
Span®61	失水山梨醇单硬脂酸酯(Crews 和 Huang,2010b)	Uniqema
Span®65	失水山梨醇三硬脂酸酯(Crews 和 Huang,2010b)	Uniqema
Span®80	失水山梨醇单油酸酯(Crews 和 Huang,2010b)	Uniqema
Span®85	失水山梨醇三油酸酯(Crews 和 Huang,2010b)	Uniqema
Tween®20	失水山梨醇月桂酸酯(Crews 和 Huang,2010b)	Uniqema
Tween®21	失水山梨醇月桂酸酯(Crews 和 Huang,2010b)	Uniqema
Tween®40	失水山梨醇单棕酸酯(Crews 和 Huang,2010b)	Uniqema
Tween®60	失水山梨醇单硬脂酸酯(Crews 和 Huang,2010b)	Uniqema
Tween®61	失水山梨醇单硬脂酸酯(Crews 和 Huang,2010b)	Uniqema
Tween®65	失水山梨醇三硬脂酸酯(Crews 和 Huang,2010b)	Uniqema
Tween®81	失水山梨醇油酸酯(Crews 和 Huang,2010b)	Uniqema
Tween®85	失水山梨醇油酸酯(Crews 和 Huang,2010b)	Uniqema
VES – STA 1	凝胶稳定剂(Crews 和 Huang,2010a,2010b)	Baker Oil Tools
Wellguard™ 7137	卤间化合物凝胶破胶剂(Carpenter,2007,2009)	Albemarle Corp.
WG – 3L VES – AROMOX® APA – T	黏弹性表面活性剂(Crews 和 Huang,2010a)	Akzo Nobel

参 考 文 献

Acker,D. B. ,Malekahmadi,F. ,2001. Delayed release breakers in gelled hydrocarbons. US Patent6 187 720,13 February 2001.

Ahlgren,J. A. ,1993. Enzymatic hydrolysis of xanthan gum at elevated temperatures andsalt concentrations. In：Proceedings Volume. Sixth Institute of Gas Technology Gas,Oil,& Environmental Biotechnology International Symposium,Colorado Springs,CO. ,29November – 1 December 1993.

API Standard RP 13B – 1,2009. Recommended practice for field testing water – based drilling fluids. API Standard API RP 13B – 1,American Petroleum Institute,Washington,DC.

Armstrong,C. D. ,2012. Compositions useful for the hydrolysis of guar in high pH environmentsand methods related thereto. US Patent Application 20120111568,10 May 2012. < http：//www. freepatentsonline. com/20120111568. html >.

Barati,R. ,Johnson,S. J. ,McCool,S. ,Green,D. W. ,Willhite,G. P. ,Liang,J. – T. ,2011. Fracturingfluid cleanup by controlled release of enzymes from polyelectrolyte complex nanoparticles. J. Appl. Polym. Sci. 121(3),1292 – 1298. http：//dx. doi. org/10. 1002/app. 33343.

Barati,R. ,Johnson,S. J. ,McCool,S. ,Green,D. W. ,Willhite,G. P. ,Liang,J. – T. ,2012. Polyelectrolyte complex nanoparticles for protection and delayed release of enzymes in alkaline pH and atelevated temperature during hydraulic fracturing of oil wells. J. Appl. Polym. Sci. 126(2),587 – 592. http：//dx. doi. org/10. 1002/app. 36845.

Bielewicz,V. D. ,Kraj,L. ,1998. Laboratory data on the effectivity of chemical breakers in mudandfiltercake(Untersuchungen zur Effektivität von Degradationsmitteln in Spülungen). ErdölErdgasKohle 114(2),76 – 79.

Boles,J. L. ,Metcalf,A. S. ,Dawson,J. C. ,1996. Coated breaker for crosslinked acid. US Patent5 497 830,assigned to BJ Services Co. ,12 March 1996.

Brannon, H. D., Tjon-Joe-Pin, R. M., 1994. Biotechnological breakthrough improves performanceof moderate to high-temperature fracturing applications. In: Proceedings Volume. 69th Annual SPE Technical Conference, vol. 1. New Orleans, 25–28 September 1994, pp. 515–530.

Cantu, L. A., Boyd, P. A., 1989. Laboratory and field evaluation of a combined fluid-loss controladditive and gel breaker for fracturing fluids. In: Proceedings Volume. SPE Oilfield ChemistryInternational Symposium, Houston, 8–10 February 1989, pp. 7–16.

Cantu, L. A., McBride, E. F., Osborne, M., 1990a. Formation fracturing process. EP Patent 401 431, assigned to Conoco Inc. and Du Pont De Nemours & Co., 12 December 1990.

Cantu, L. A., McBride, E. F., Osborne, M., 1990b. Well treatment process. EP Patent 404 489, assigned to Conoco Inc. and Du Pont De Nemours & Co., 27 December 1990.

Carpenter, J. F., 2007. Breaker composition and process. US Patent 7 223 719, assigned to Albemarle Corporation, Richmond, VA, 29 May 2007. < http://www.freepatentsonline.com/7223719.html >.

Carpenter, J. F., 2009. Bromine-based sulfamate stabilized breaker composition and process. USPatent 7 576 041, assigned to Albemarle Corporation, Baton Rouge, LA, 18 August 2009. < http://www.freepatentsonline.com/7576041.html >.

Cordova, M., Cheng, M., Trejo, J., Johnson, S. J., Willhite, G. P., Liang, J.-T., Berkland, C., 2008. Delayed HPAM gelation via transient sequestration of chromium in polyelectrolyte complexnanoparticles. Macromolecules 41(12), 4398–4404. http://dx.doi.org/10.1021/ma800211d.

Craig, D., 1991. The degradation of hydroxypropyl guar fracturing fluids by enzyme, oxidative, and catalyzed oxidative breakers. Ph. D. Thesis, Texas A & M University.

Craig, D., Holditch, S. A., 1993a. The degradation of hydroxypropyl guar fracturing fluids byenzyme, oxidative, and catalyzed oxidative breakers: Pt. 1: linear hydroxypropyl guar solutions: topical report, February 1991–December 1991. Gas. Res. Inst. Rep. GRI-93/04191, Gas. Res. Inst.

Craig, D., Holditch, S. A., 1993b. The degradation of hydroxypropyl guar fracturing fluids byenzyme, oxidative, and catalyzed oxidative breakers: Pt. 2: crosslinkedhydroxypropyl guargels: topical report, January 1992–April 1992. Gas. Res. Inst. Rep. GRI-93/04192, Gas. Res. Inst.

Craig, D., Holditch, S. A., Howard, B., 1992. The degradation of hydroxypropyl guar fracturingfluids by enzyme, oxidative, and catalyzed oxidative breakers. In: Proceedings Volume. 39th Annual Southwestern Petroleum Short Course Association Inc. et. al Meeting, Lubbock, TX, 22–23 April 1992, pp. 1–19.

Crews, J. B., 2006. Bacteria-based and enzyme-based mechanisms and products for viscosityreduction breaking of viscoelastic fluids. US Patent 7 052 901, assigned to Baker HughesIncorporated, Houston, TX, 30 May2006. < http://www.freepatentsonline.com/7052901.html >.

Crews, J. B., 2007a. Aminocarboxylic acid breaker compositions for fracturing fluids. USPatent 7 208 529, assigned to Baker Hughes Incorporated Houston, TX, 24 April 2007. < http://www.freepatentsonline.com/7208529.html >.

Crews, J. B., 2007b. Polyols for breaking of fracturing fluid. US Patent 7 160 842, assigned to Baker Hughes Incorporated, Houston, TX, 9 January 2007. < http://www.freepatentsonline.com/7160842.html >.

Crews, J. B., 2010. Saponified fatty acids as breakers for viscoelastic surfactant-gelled fluids. US Patent 7 728 044, assigned to Baker Hughes Incorporated, Houston, TX, 1 June 2010. < http://www.freepatentsonline.com/7728044.html >.

Crews, J. B., Huang, T., 2010a. Unsaturated fatty acids and mineral oils as internal breakers forves-gelled fluids. US Patent 7 696 134, assigned to Baker Hughes Incorporated, Houston, TX, 13 April 2010. < http://www.freepatentsonline.com/7696134.html >.

Crews, J. B., Huang, T., 2010b. Use of oil-soluble surfactants as breaker enhancers for ves-gelledfluids. US Patent

7 696 135, assigned to Baker Hughes Incorporated, Houston, TX, 13 April 2010. < http://www. freepatentsonline. com/7696135. html >.

Dawson, J. C. , Le, H. V. , 1995. Controlled degradation of polymer based aqueous gels. US Patent5 447 199, assigned to BJ Services Co. ,5 September 1995.

Fodge, D. W. , Anderson, D. M. , Pettey, T. M. , 1996. Hemicellulase active at extremes of pH andtemperature and utilizing the enzyme in oil wells. US Patent 5 551 515, assigned to Chemgen Corp. , 3 September 1996.

Gulbis, J. , King, M. T. , Hawkins, G. W. , Brannon, H. D. , 1990a. Encapsulated breaker for aqueouspolymeric fluids. In: Proceedings Volume. 9th SPE Formation Damage Control Symposium, Lafayette, LA, 22 – 23 February 1990, pp. 245 – 254.

Gulbis, J. , Williamson, T. D. A. , King, M. T. , Constien, V. G. , 1990b. Method of controlling release ofencapsulated breakers. EP Patent 404 211, assigned to Pumptech NV and Dowell SchlumbergerSA, 27 December 1990.

Gulbis, J. , King, M. T. , Hawkins, G. W. , Brannon, H. D. , 1992. Encapsulated breaker for aqueouspolymeric fluids. SPE Prod. Eng. 7(1) ,9 – 14.

Gupta, D. V. S. , Cooney, A. , 1992. Encapsulations for treating subterranean formations and methodsfor the use thereof. WO Patent 9 210 640, assigned to Western Co. , North America, 25 June1992.

Gupta, D. V. S. , Prasek, B. B. , 1995. Method for fracturing subterranean formations using controlledrelease breakers and compositions useful therein. US Patent 5 437 331, assigned to Western.

Co. , North America, 1 August 1995. Harms, W. M. , 1992. Catalyst for breaker system for high viscosity fluids. US Patent 5 143 157, assigned to Halliburton Co. , 1 September 1992.

Harris, R. E. , Hodgson, R. J. , 1998. Delayed acid for gel breaking. US Patent 5 813 466, assignedtoCleansorb Limited, GB, 29 September 1998. < http://www. freepatentsonline. com/5813466. html >.

Hawkins, G. W. , 1986. Molecular weight reduction and physical consequences of chemicaldegradation of hydroxypropylguar in aqueous brine solutions. In: Proceedings 192nd ACSNational Meetings American Chemical Society Polymeric Materials Science EngineeringDivision of Technology Program, vol. 55. Anaheim, Calif, 7 – 12 September1986, pp. 588 – 593.

Horton, R. L. , Prasek, B. , Growcock, F. B. , Kippie, D. , Vian, J. W. , Abdur – Rahman, K. B. , Arvie, Jr. , M. , 2007. Surfactant – polymer compositions for enhancing the stability of viscoelastic – surfactant based fluid. US Patent 7 157 409, assigned to M – I LLC, Houston, TX, 2 January2007. < http://www. freepatentsonline. com/7157409. html >.

Horton, R. L. , Prasek, B. B. , Growcock, F. B. , Kippie, D. P. , Vian, J. W. , Abdur – Rahman, K. B. , Arvie, M. , 2009. Surfactant – polymer compositions for enhancing the stability of viscoelastic – surfactant based fluid. US Patent 7 517 835, assigned to M – I LLC, Houston, TX, April 14 2009. < http://www. freepatentsonline. com/7517835. html >.

Hunt, C. V. , Powell, R. J. , Carter, M. L. , Pelley, S. D. , Norman, L. R. , 1997. Encapsulated enzymebreaker and method for use in treating subterranean formations. US Patent 5 604 186, assignedto Halliburton Co. , 18 February 1997.

Jihua, C. , Sui, G. , 2011. Rheological behaviors of bio – degradable drilling fluids in horizontal drillingof unconsolidated coal seams. Int. J. Info. Technol. Comput. Sci. 3(3) ,1 – 7.

Johnson, S. , Barati, R. , McCool, S. , Green, D. W. , Willhite, G. P. , Liang, J. – T. , 2011. Polyelectrolyte complex nanoparticles to entrap enzymes for hydraulic fracturing fluid cleanup. PreprintsAmer. Chem. Soc. Div. Pet. Chem. 56(2) , 168 – 172. < http://pubs. acs. org/cgi – bin/preprints/display? div = petr&meet = 242&page = 88718. pdf >.

King, M. T. , Gulbis, J. , Hawkins, G. W. , Brannon, H. D. , 1990. Encapsulated breaker for aqueous polymeric fluids. In: Proceedings Volume. Canadian Institute of Mining, Metallurgy and Petroleum Society/SPE International Technology Meeting, vol. 2. Calgary, Canada, 10 – 13 June 1990.

Kyaw, A., Binti, NorAzahar, B. S., Tunio, S. Q., 2012. Fracturing fluid(guar polymer gel)degradation study by using oxidative and enzyme breaker. Res. J. Appl. Sci. Eng. Technol. 4 (12), 1667 − 1671. < http://maxwellsci.com/print/rjaset/v4 − 1667 − 1671. pdf >.

Langemeier, P. W., Phelps, M. A., Morgan, M. E., 1989. Method for reducing the viscosity of aqueousfluids. EP Patent 330 489, 30 August 1989. Laramay, S. B., Powell, R. J., Pelley, S. D., 1995. Perphosphate viscosity breakers in well fracturefluids. US Patent 5 386 874, assigned to Halliburton Co., 7 February 1995.

Ma, Y., Xue, Y., Dou, Y., Xu, Z., Tao, W., Zhou, P., 2004. Characterization and gene cloningof a novel/? − mannanase from alkaliphilic bacillus sp. n16 − 5. Extremophiles 8(6), 447 − 454. http://dx.doi.org/10.1007/s00792 − 004 − 0405 − 4.

Manalastas, P. V., Drake, E. N., Kresge, E. N., Thaler, W. A., McDougall, L. A., Newlove, J. C., Swarup, V., Geiger, A. J., 1992. Breaker chemical encapsulated with a crosslinked elastomercoating. US Patent 5 110 486, assigned to Exxon Research & Eng. Co., 5 May 1992.

McDougall, L. A., Malekahmadi, F., Williams, D. A., 1993. Method of fracturing formations. EPPatent 540 204, assigned to Exxon Chemical Patents In, 5 May 1993.

Mitchell, T. O., Card, R. J., Gomtsyan, A., 2001. Cleanup additive. US Patent 6 242 390, assignedto Schlumberger Technol. Corp., 5 June 2001.

Mondshine, T. C., 1993. Process for decomposing polysaccharides in alkaline aqueous systems. USPatent 5 253 711, assigned to Texas United Chemical Corp., Houston, TX, 19 October 1993. < http://www.freepatentsonline.com/5253711.html >.

Muir, D. J., Irwin, M. J., 1999. Encapsulated breakers, compositions and methods of use. WO Patent9 961 747, assigned to 3M Innovative Propertie C, 2 December 1999.

Noran, L., Vitthal, S., Terracina, J., 1995. New breaker technology for fracturing high − permeability formations. In: Proceedings Volume. SPE Europe Formation Damage Control Conference, TheHague, Neth, 15 − 16 May 1995, pp. 187 − 199.

Norman, L. R., Laramay, S. B., 1994. Encapsulated breakers and method for use in treating subterranean formations. US Patent 5 373 901, assigned to Halliburton Co., 20 December 1994.

Norman, L. R., Turton, R., Bhatia, A. L., 2001. Breaking fracturing fluid in subterranean formation. EP Patent 1 152 121, assigned to Halliburton Energy Serv., 7 November 2001.

Prasek, B. B., 1996. Interactions between fracturing fluid additives and currently used enzyme breakers. In: Proceedings Volume. 43rd Annual Southwestern Petroleum Short CourseAssociation Inc. et al. Meeting Lubbock, Texas, 17 − 18 April 1996, pp. 265 − 279.

Sarwar, M. U., Cawiezel, K., Nasr − El − Din, H., 2011. Gel degradation studies of oxidative andenzyme breakers to optimize breaker type and concentration for effective break profiles at lowand medium temperature ranges. In: Proceedings of SPE Hydraulic Fracturing Technology Conference. No. 140520 − MS. SPE Hydraulic Fracturing Technology Conference, 24 − 26January 2011, The Woodlands, Texas, USA, Society of Petroleum Engineers, Dallas, Texas, pp. 1 − 21. http://dx.doi.org/10.2118/140520 − MS.

Shuchart, C. E., Terracina, J. M., Slabaugh, B. F., McCabe, M. A., 1999. Method of treating subterranean formation. EP Patent 916 806, assigned to Halliburton Energy Serv., 19 May1999.

Slodki, M. E., Cadmus, M. C., 1991. High − temperature, salt − tolerant enzymic breaker of xanthangum viscosity. In: Donaldson, E. C. (Ed.), Microbial Enhancement of Oil Recovery: RecentAdvances: Proceedings of the 1990 International Conference on Microbial Enhancement of OilRecovery of Developments in Petroleum Science, vol. 31. Elsevier Science Ltd., pp. 247 − 255.

Swarup, V., Peiffer, D. G., Gorbaty, M. L., 1996. Encapsulated breaker chemical. US Patent5 580 844, assigned to

Exxon Research & Eng. Co. ,3 December 1996.

Syrinek, A. R. , Lyon, L. B. , 1989. Low temperature breakers for gelled fracturing fluids. US Patent4 795 574, assigned to Nalco Chemical Co. ,3 January 1989.

Tjon – Joe – Pin, R. M. ,1993. Enzyme breaker for galactomannan based fracturing fluid. USPatent 5 201 370, assigned to BJ Services Company, Houston, TX, 13 April 1993. < http://www. freepatentsonline. com/5201370. html >.

Walker, M. L. , Shuchart, C. E. , 1995. Method for breaking stabilized viscosified fluids. US Patent5 413 178, assigned to Halliburton Co. ,9 May 1995.

Williams, M. M. , Phelps, M. A. , Zody, G. M. , 1987. Reduction of viscosity of aqueous fluids. EPPatent 222 615, 20 May 1987.

17 杀 菌 剂

压裂施工的用水一般为浅层地表水或湖泊及河流中水,这些水中含有大量细菌,水中的溶解氧会增加细菌的滋生从而破坏水质及压裂液性能。压裂液中常常添加除氧剂或杀菌剂以保证压裂液性能稳定。

然而常用的杀菌剂与除氧剂在压裂液体系中会发生负反应,从而抑制其单剂的作用或影响其他添加剂的作用,从而影响压裂液体系整体性能。McCurdy 于 2008 年对各种杀菌剂与压裂液中除氧剂的相互作用与影响进行了详细的研究与介绍(McCurdy,2008)。

石油行业内应用的流体普遍存在的一个问题是,储层内含有的生物细菌会在地层内生成大量的硫化氢(H_2S)。硫化氢会腐蚀井内金属设备,生成硫化铁,这种现象的存在大大提高了现场作业成本,同时也会对环境造成伤害。

一些严重的腐蚀也可能是由于特定细菌生物与酸作用而产生的。这些细菌生物往往是硫酸盐还原菌在厌氧环境下生成的。一旦生成了这种硫酸盐还原菌,那么很难恢复液体体系的生物菌群环境,并很难进行生物学的控制。

当在金属表面地层形成细菌膜,会严重腐蚀石油生产设施,这会导致这些石油装备性能严重退化从而影响生产。

据估计,由微生物腐蚀导致的设备故障占现场故障率的15%~30%。

因此,应当采取有效的控制细菌措施来消除这些不利的影响。目前行业内已经开发了多种杀菌剂和无菌技术来控制细菌腐蚀,这些措施被证明是有效的。同时也开发了相应的细菌检测技术用于现场的细菌控制及检测。

17.1 生长机制

17.1.1 油田化学剂滋生的细菌

细菌生长实验通常是将含有几种常用化学处理剂的水注入油井中进行研究的。研究表明,某些化学品可以作为细菌生长的氮源、磷源以及碳源(Sunde 等,1990)。因此,水处理添加剂对细菌滋生的作用是巨大的,应在添加剂优选前考虑到细菌的影响作用。

在其他的实验中发现油田用水中分离出来的硫酸盐还原菌比实验室用标准细菌培养出来的硫酸盐还原菌数量要大得多。不同化学试剂对硫酸盐还原菌的刺激作用差别较大,其不同主要取决于物质分类、活性以及与细菌的适应性。

硫酸盐还原菌的选择性培养可用于抵抗一些有毒化合物。因此,在井底及地层内使用杀菌剂可完全抑制地层中的硫酸盐还原细菌的生命活性,实际杀菌剂的使用剂量远比实验室中得到的数据要高得多(Kriel 等,1993)。

研究表明,硫酸盐还原菌可通过油井返排水的重新利用而被再次注入地层,在地层高温下恢复其生物活性(Salanitro 等,1993)。通过降解地层水中的脂肪酸,这些生物可以在较冷的区域生存并维持缓慢增长。

17.1.2 数学模型

可通过数学模型研究油藏酸化过程中硫酸盐还原菌的生长。该模型是基于守恒方程的一维数值模型,其中的影响因素包括细菌的生长率、运动、营养成分、混合水以及地层中 H_2S 的吸附等。通过现场数据分析微生物生成的硫化氢数据如下(Sunde 等,1993):

(1)在地层水和注入混合区域产生的 H_2S(混合区模型);

(2)接近注入井储层岩石上硫酸盐还原菌产生的 H_2S(生物模型)。

Sunde 等 1993 年使用生物模型分析了 Gullfaks 油田 3 口生产井产生的 H_2S 数据,而数据表明现场数据不适用于混合区模型进行分析。

菌落生长模型:

实验室内进行了大量相关实验,通过加入不同量的硫酸铜研究细菌生长随时间的变化关系,具体结果见图 17.1。这个测试需要衡量菌落直径。

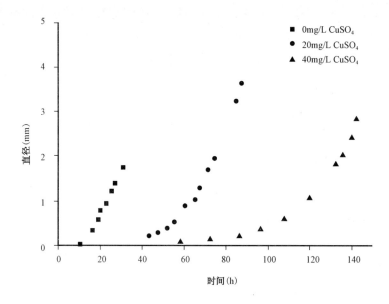

图 17.1 硫酸铜对黏质沙雷氏菌菌落生长的影响(Rodin 等,2005)

Rodin 等于 2005 年得出了一个简化的菌落生长模型。根据该模型,在生长过程中,菌落依次经历了指数和线性增长阶段。然而,排除营养物质浓度的影响,菌落对数生长期持续时间是有限的。在指数生长阶段,菌落直径变化过程用公式(17.1)描述:

$$d = d_0 \exp\left(\frac{\mu'_m}{2} t\right) \tag{17.1}$$

式中,d 为菌落直径;d_0 为单个细胞的有效直径;t 为时间;μ'_m 为最大生长率。

相反,线性增长阶段发生在营养受限期,从时间 t_1 开始,然后以恒定速率增加菌落直径。根据式(17.2)可得到 k_d:

$$d = d_{t_1} + k_d(t - t_1) \tag{17.2}$$

结合式(17.1)和式(17.2),可以得到公式(17.3):

$$t = \frac{d}{k_d} + \frac{2}{\mu'_m}\left[\ln\left(\frac{2k_d}{d_0\mu'_m}\right) - 1\right] \tag{17.3}$$

式(17.3)中计算了与可见的大菌落线性增长阶段及不可见的菌落指数生长期的关键参数。此属性大大简化了 μ_m 参数的实验测定过程(Rodin 等,2005)。

17.1.3 硫酸盐还原菌

硫酸盐还原菌属于可自生长类细菌(Barton 和 Fauque,2009),已知的硫酸盐还原菌有 60 属 220 种。所有的硫酸盐还原菌都使用硫酸盐作为终端电子载体。在这种方式中,构成了一个独特的微生物生理群,以厌氧方式进行三磷酸腺苷(ATP)合成。新的硫酸盐还原菌种类不断被发现(Miranda – Tello 等,2003;Youssef 等,2009;Agrawal 等,2010),例如 2003 年,在墨西哥油田的一个分离器中得到了一种新的硫酸盐还原菌 MET2T(Miranda – Tello 等,2003)

这些细菌可以使用各种各样的化合物作为电子供体。被氧化还原的金属族群为蛋白质代谢提供基础。特别是,其行为作用于可溶性电子转移蛋白质并通过跨膜影响氧化还原复合物。利用纯培养和培养基质中硫酸盐还原菌对碳氢化合物的作用原理,可以去除土壤中芳香族碳氢化合物污染。

一些硫酸盐还原菌甚至可以用于减少氯化物。例如 3 – 氯苯甲酸、氯乙烯和硝基化合物,硫酸盐还原菌还可以减少一些重金属污染。因此,研究人员已经开始实验研究硫酸盐还原菌对土壤中有毒金属的反应及菌株的生物修复作用。

硫酸盐还原菌的代谢过程会产生大量的硫化氢。由此产生的硫化氢气体溶解在水中使水变成酸性,这将给石油设备带来严重的腐蚀和破坏。

17.1.4 细菌腐蚀

细菌腐蚀通常被认为是微生物活动导致的腐蚀。微生物的腐蚀作用涉及微生物对腐蚀作用的启动和加速两个方面。微生物的代谢产物几乎对大多数工程材料都会产生影响,但对常用的耐腐蚀合金不发生作用,如不锈钢。

微生物对腐蚀的影响被严重低估了。因为大多数微生物引起的腐蚀一般发生在局部,且以点蚀为主。一般来说,这种类型的腐蚀结果,在腐蚀失重率变化上数据相对较低,还会导致电阻的变化,改变受影响的总面积。这使得传统的腐蚀性能测试对微生物腐蚀的影响评估较为困难(Pope 等,1992)。

17.1.4.1 pH 值调节

细菌的代谢产物一般为弱酸。硫酸盐还原菌的活性取决于潜在的二次反应环境的 pH 值水平:
(1)硫酸铁沉淀;
(2)硫化物离子的氧化过程对硫代硫酸中氧的变化;
(3)代谢硫代硫酸钠或其他硫化合物。

通过这种方式可以分析和解释细菌腐蚀的开始和发展过程。

17.1.4.2 强化杀菌剂

为了有效地处理水中细菌产生的污染,采用具有快速作用杀菌剂是必要的技术手段。这需要杀菌剂具有快速作用性能,现场实施过程中需要在添加其他化学添加剂之前加入杀菌剂,并在水中快速起作用,然后在较短时间内注入井内。在某些情况下,为了获得快速杀菌作用,需要借助于杀菌剂的协同效应(Bryant 等,2009)。

季铵盐表面活性剂同时也可以是杀菌剂。例如 19N^{TM}是一种阳离子表面活性剂,同时也是一种杀菌剂。当使用如次氯酸钠或戊二醛组合的杀菌剂时,某些情况下细菌问题可在 5min 内得到解决。

虽然季铵盐表面活性剂可作为用杀菌剂一起使用,季铵盐表面活性剂与阴离子降阻剂不配伍,然而一旦到达地层内也会起到杀菌作用。

研究表明,季铵盐表面活性剂与阴离子降阻剂的不相容性主要是来源于分子内部作用,这会最终导致生成沉淀。此外,一些杀菌剂(例如氧化剂)也会降低降阻剂性能(Bryant 等,2009)。

17.2 性能控制

压裂液中细菌污染将严重影响压裂液性能。如果压裂液未经除菌处理,其中的硫酸盐还原菌和酸液产生的细菌会大量滋生。

压裂液中常含有大量聚丙烯酰胺类聚合物、多烷基糖类聚合物以及其他的一些有机化合物,这些物质的存在为细菌生长提供了食物来源。

通常压裂用水一般来自河流、湖泊以及油田废水,这其中含有大量细菌。

测试包括溴硝醇、戊二醛、戊二醛/季铵盐化合物混合物、异噻唑啉、四羟甲基硫酸酯和 2,2-二溴-3-次氮基丙酰胺等杀菌剂,结果显示四羟甲基硫酸磷表现良好,它可以快速杀死细菌,抑制细菌发酵及硫酸盐还原菌的活性。通过长期的实验发现,戊二醛及其混合物对油井内耗氧细菌及酸液产生菌的抑制作用不明显(Johnson 等,2008;Fichter 等,2009)。

得克萨斯州沃斯堡盆地的巴奈特页岩压裂中进行了杀菌剂性能测试。巴奈特页岩是地下 1.5mile❶ 深且在地下延伸 5000mile^2 的天然气储层。这个区域是美国第二大石油和天然气生产基地(Johnson 等,2008)。

17.3 杀菌剂处理措施

17.3.1 裂缝性地层

裂缝性地层内一个比较特殊的问题是由于重复压裂导致地层环境已经被细菌严重污染。在这种情况下,必须在压裂液中添加足够的杀菌剂以达到杀死地层内细菌的作用。重复压裂过程使得杀菌剂可以分布在所有液体接触到的地层表面并抑制细菌作用(McCabe 等,1991)。

17.3.2 间歇性加入杀菌剂

间歇式添加技术(Hegarty 和 Levy,1996a,1996b,1996c)如下:
(1)添加一定剂量的杀菌剂,保证杀菌剂能够快速有效地杀灭细菌。

---

❶ 1mile=1.609344km

(2)间歇性添加杀菌剂并有效控制剂量。这意味着在一段时间内要控制杀菌剂的剂量,随后一段时间内不添加或少剂量添加杀菌剂。重复这一过程,作为间歇式操作措施。

这种工艺减少并有效控制了生物杀菌剂的加量,同时减少了杀菌剂对油田生产用水质的影响。杀菌剂的加入周期一般为 2～15d,杀菌剂作用时间一般为 4～8h(Moody 和 Montgomerie,1996)。

17.3.3 非生物控制

细菌控制化学处理措施意味着更高的经济成本以及环境污染。由于有毒生物杀菌剂的监管压力越来越大,因此需要开发更符合环保要求的控制措施。

17.3.3.1 生物竞争排除技术

除了向井内添加杀菌剂,另一种方法是改善地层的生态系统。通过改变土壤中微生物种群及数量,可以控制硫酸盐还原菌的产生,并降低硫酸盐还原菌的浓度。

该技术采用了低浓度的水溶性营养液,选择性地培养土壤中微生物种群的生长,从而抑制可产生 H_2S 的硫酸盐还原细菌的生成(Sandbeck 和 Hitzman,1995;Hitzman 和 Dennis,1997)。

17.3.3.2 细菌膜抑制剂

用季铵盐添加剂进行实验测试的数据显示出低的表面点蚀和腐蚀速率(Enzien 等,1996)。季铵盐试验测试在压裂液中的杀菌效果似乎是最小的。这些结果表明,季铵盐可以有效防止微生物的腐蚀影响,与其他杀菌剂相比季铵盐可持续有效地防止细菌在表面的繁殖及腐蚀影响。

17.3.3.3 离子强度的周期性变化

为了有效控制微生物的影响,应该考虑细菌的形成及影响其生态因子的原理。细菌的生命活动过程始于对周围岩心的适应性及吸附作用。单纯的硫酸盐还原菌在原油培养液中是不活跃的。

在油气藏开发过程中,油气的开采完全取决于烃氧化菌的作用,这是石油生成的主要原因。如果在储层生态条件下,微生物的形成过程被改变,所建立的食物链被破坏,微生物活性及生长会停止。实验结果表明在矿化度显著不同情况下,向地层内周期性注入水,同时考虑微生物形成的生态特征,可以有效控制油藏中生物环境,从而在对环境无干扰条件下控制井内有害菌的影响(Blagov 等,1990)。

17.4 特殊化学品

多年来,不同杀菌技术已成功应用于大多数油田的水处理措施。这其中包括氧化剂如氯和溴的产品,非氧化性杀菌剂包括异噻唑啉酮、季铵化合物、有机溴化物和戊二醛。

通过实验已经评估测试了多种用于页岩压裂液中抑制硫酸盐还原菌的生物杀菌剂的有效性(Struchtemeyer 等,2012)。实验测试的杀菌剂包括四(羟甲基)硫酸盐、次氯酸钠、二癸基二甲基氯化铵、3-N-丁基十四烷基氯化磷及戊二醛。测试浮游细胞和生物的最小抑制浓度。结果表明,硫酸盐还原菌的负面影响导致生物膜的形成对杀菌剂的性能有影响。

杀菌剂的错误使应用现象在石油工业内普遍存在。杀菌剂的滥用一般是在应用前未对杀菌剂进行足够的研究及分析。Boivin 在 1994 年对杀菌剂的选择指导方法进行了文献综述(Boivin,

1994)。微生物的早期检测是重要的,对微生物的修复行动必须尽快地采取处理措施。

这些措施应防范对环境的退化影响。在一般情况下,杀菌剂的作用是控制体系内的细菌活性。然而,单独使用杀菌剂通常不会解决生物问题。下面介绍了杀菌剂选择的五项基本要求(Zhou,1990):

(1)广泛的杀菌能力及耐腐蚀性能;
(2)好的抑制能力,方便运输和使用;
(3)低毒或无毒特性,不对人类造成伤害,符合环境管理条例;
(4)良好的相容性,不损坏或干扰钻井液或其他化学试剂性能;
(5)细菌杀灭效果不会受到细菌的环境适应性影响。

水基压裂液中含有瓜尔胶或其他天然聚合物,可以通过添加杂环硫化合物以防止细菌的影响。这种稳定方法防止了细菌不必要地降解压裂液,如压裂液流变性能的降低(这是进行水力压裂操作所必需的)。表17.1、图17.2 和图17.3 中列举了一些适用于压裂液的杀菌剂。

表 17.1 杀菌剂(Kanda 等,1988;Kanda 和 Kawamura,1989)

名称	名称
巯基苯并咪唑①	1,3,4-噻二唑-2,5-二硫醇①②
2-巯基苯并噻唑	2-巯基苯并噁唑
2-巯基苯垪	2-巯基噻唑啉
2-硫代咪唑啉酮	2-硫醇硫基咪唑啉
4-噻唑烷-2-酮基硫醇	N-巯基吡啶-2-硫醇

① 用于瓜尔胶。
② 用于黄胞胶。

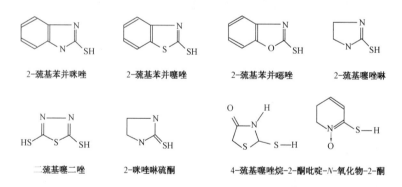

图 17.2 水基压裂液中应用的杀菌剂

图 17.3 硫醇

参 考 文 献

Agrawal, A., Vanbroekhoven, K., Lal, B., 2010. Diversity of culturable sulfidogenic bacteria intwo oil – water separation tanks in the north – eastern oil fields of india. Anaerobe 16(1), 12 – 18. http://www.sciencedirect.com/science/article/B6W9T – 4W7J150 – 1/2/bd2cb4d4c087315b0ed914986b4ed3c2.

Barton, L. L., Fauque, G. D., 2009. Biochemistry, physiology and biotechnology of sulfate – reducing bacteria. In: Laskin, A. I., Sariaslani, S., Gadd, G. M. (Eds.), Advances in Applied Microbiology, vol. 68. Academic Press, pp. 41 – 98 (Chapter 2). <http://www.sciencedirect.com/science/article/B7CSY – 4W79HR5 – 4/2/1b6c4d955860bf7b7eecc73674593b03>.

Blagov, A. V., Prazdnikova, Z. F., Praporshchikov, V. I., 1990. Use of ecological factors for controlling biogenic sulfate reduction. NeftKhoz 5, 48 – 50.

Boivin, J., 1994. Oil industry biocides. Mater. Perf. 34(2), 65 – 68.

Bryant, J. E., McMechan, D. E., McCabe, M. A., Wilson, J. M., King, K. L., 2009. Treatment fluidshaving biocide and friction reducing properties and associated methods. US Patent Application20 090 229 827, 17 September 2009. <http://www.freepatentsonline.com/20090229827.html>.

Enzien, M. V., Pope, D. H., Wu, M. M., Frank, J., 1996. Nonbiocidal control of microbiologically influenced corrosion using organic film – forming inhibitors. In: Proceedings Volume. 51st Annual NACE International Corrosion Conference(Corrosion 96) Denver, 24 – 29 March 1996.

Fichter, J. K., Johnson, K., French, K., Oden, R., 2009. Biocides control barnett shale fracturing fluid contamination. Oil Gas J. 107(19), 38 – 44.

Hegarty, B. M., Levy, R., 1996a. Control of oilfield biofouling. CA Patent 2 160 305, 13 April 1996.

Hegarty, B. M., Levy, R., 1996b. Control of oilfield biofouling. EP Patent 706 759, 17 April 1996.

Hegarty, B. M., Levy, R., 1996c. Procedure for combatting biological contamination in petroleum production (procede pour combattrel'encrassementbiologiquedans la production de petrole). FR Patent 2 725 754, 19 April 1996.

Hitzman, D. O., Dennis, D. M., 1997. Sulfide removal and prevention in gas wells. In: ProceedingsVolume. SPE Production and Operations Symposium, Oklahoma City, 9 – 11 March 1997, pp. 433 – 438.

Johnson, K., French, K., Fichter, J. K., Oden, R., 2008. Use of microbiocides in Barnett shalegas well fracturing fluids to control bacteria related problems. In: Corrosion, 2008. NACEInternational, New Orleans, LA.

Kanda, S., Kawamura, Z., 1989. Stabilization of xanthan gum in aqueous solution. US Patent4 810 786, 7 March 1989.

Kanda, S., Yanagita, M., Sekimoto, Y., 1988. Stabilized fracturing fluid and method of stabilizing fracturing fluid. US Patent 4 721 577, 26 January 1988.

Kriel, B. G., Crews, A. B., Burger, E. D., Vanderwende, E., Hitzman, D. O., 1993. The efficacy offormaldehyde for the control of biogenic sulfide production in porous media. In: ProceedingsVolume. SPE Oilfield Chemistry International Symposium, New Orleans, 2 – 5 March 1993, pp. 441 – 448.

18 支 撑 剂

一般来说,支撑剂颗粒应该悬浮在压裂液中,以避免在压裂施工结束,压力扩散后裂缝完全闭合,从而形成导流通道,以确保碳氢化合物可以从地层内流出。至少形成一条人工裂缝,并且有部分支撑剂基本到位,压裂液就可以降低黏度了,从而从地层中返排出来(Todd 等,2006)。

18.1 液体滤失

在某些情况下,压裂施工过程中会滤失部分压裂液,例如,支撑剂随着压裂液漏失到地层中的天然裂缝中。这是有问题的,因为天然裂缝往往比人工裂缝的应力更高。这些高应力可能会损坏支撑剂,使其在天然裂缝形成无导流的段塞,这样会阻止烃类流过天然裂缝。

传统上,作业者试图通过在压裂中使用降滤失剂来解决这个问题。常规的降滤失剂一般包括球状的刚性粒子。这些添加剂的使用是有问题的,因为这样的添加剂可能需要有不同粒径分布以实现高效控制液体滤失。

例如,当使用这种添加剂堵塞地层中的孔喉时,需要足够的相对较大粒径颗粒堵塞孔喉的大部分。还需要足够量的较小粒径的颗粒,以阻挡大颗粒之间的孔隙。而且,某些常规的流体滤失控制添加剂,可能很难获得所需的这种粒径分布,需要对材料进行再加工,从而增加费用,例如,可以通过低温研磨来获得所需的粒径分布(Todd 等,2006)。

18.2 示踪剂

最初开发的压裂示踪剂是用来判断支撑剂位置、压裂液返排和支撑带清理的(Asadi 等,2008)。

已经可以用非放射性化学示踪剂评价这些行为。可以在压裂液的不同阶段加入几类化学示踪剂进行监测。用这种方式,可以评估每段流体的返排效率。

假定在多阶段的施工过程中,每段压裂的返排是可以监测的。从这些数据中可以确定支撑剂径向铺置的有效性和垂向沟通小层的情况。

用流变测量的方法研究了化学示踪剂的配伍性。选择用锆交联羧甲基羟丙基瓜尔胶和硼酸盐交联瓜尔胶压裂液作为研究对象(Sullivan 等,2004)。

研发了具有设计条件、设计变量、裂缝几何尺寸、生产模块和成本模块模型的集成优化工具模型。该集成模型可以用于低渗透油藏(Asadi 等,2008)。

18.3 支撑剂的成岩作用

已经确定了造成裂缝导流能力降低的许多机理,包括支撑剂颗粒的机械破碎、地层细小颗粒堵塞、支撑剂嵌入、地层剥落、压裂液伤害、应力循环交变伤害、沥青质沉积伤害或支撑剂的溶解(duenckel 等,2012)。这些因素组合以后,与通常公布的标准条件下测量的导流能力数据相比,可以成数量级地降低有效导流能力。

另一种机制是假定支撑剂导流能力的降低随着时间而变化。这种机制被称为成岩作用，类似于溶解和再沉淀的过程，会降低孔隙度、渗透率和在沉积过程中的支撑带的强度。

成岩作用的发生和程度是由闭合应力、储层温度、支撑剂类型和地层岩石矿物学所控制的。此外，仍有一些不确定的无法预测的成岩作用发生。

已有相关以下研究成果报告，包括高温静态试验、在油藏条件下的长期测试、析出相的详细分析、环境对支撑剂力学性能的影响、各种储层页岩的化学和矿物学分析以及实际生产井中返出的支撑剂样品的评价（Duenckel 等，2012）。

在支撑剂的表面可以形成结晶析出物。在所有类型的支撑剂上都会出现这一问题，包括陶粒、石英砂、树脂涂层材料，甚至在惰性金属球或玻璃上。

在没有氧化铝时，可以形成沸石。在精确模拟地层条件下，用实际储层页岩岩心样品进行试验表明，如果成岩沉淀确实发生，会显著影响评价条件下的裂缝导流能力。此外，其他油气藏条件会自然会防止沸石的形成。这项工作的结果将有助于增产措施工程师进行支撑剂选择和施工设计。虽然有许多机制描述支撑剂性能的下降，这些研究表明，沸石沉淀不是大多数井作业要担心的主要问题（Duenckel 等，2012）。

18.4 支撑剂

为了更好地进行油气井增产措施，最好的支撑剂和液体必须与良好的设计方案和合适的设备相结合。支撑剂的选择是决定增产措施成败的重要因素。为了每口井都可以选择最好的支撑剂，有必要了解支撑剂的性能。

支撑剂应该具有在不同地层压力下的高渗透、高抗压缩、低密度和良好的耐酸蚀性能。部分支撑剂在表 18.1 中列出。

表 18.1 支撑剂的基本性能

材料	描述/性能	参考文献
铝矾土	标准的	Andrews（1987）和 Fitzgibbon（1986）
铝矾土 + ZrO_2	抗高应力	Khaund（1987b）
砂子	高压时渗透率较低	
轻质	控制密度	Bienvenu（1996）
陶粒	可以制成球状的	Gibb 等（1990）
黏土		Fitzgibbon（1988，1989）和 Khaund（1987a）

18.4.1 石英砂

石英砂是最简单的支撑剂材料。石英砂虽然便宜，但在更高应力条件下，其渗透率降低严重。

18.4.2 陶粒

烧结陶粒早已用做井的支撑剂（Laird 和 Beck，1989）。每个颗粒都是由包括矿物颗粒、碳化硅和黏合剂组成的原材料的核芯组成。该混合物包括化学束缚水或硫矿物，在加热过程中充分混合。

核心有大量的密闭的空气泡。每个球体都有围绕核心的包括氧化铝和氧化氧化镁的金属氧化物外壳。烧制的陶粒密度小于2.2g/cm³。

18.4.3 铝矾土

含有二氧化硅的烧结铝矾土小球是标准的支撑剂材料。粒径范围0.02～0.3μm，烧结前，在混合物中加入2%氧化锆可以提高支撑剂的强度。以下给出了支撑剂生产的主要流程。

制备18.1：将极细的颗粒从天然铝矾土中分离出来。细粉末是非煅烧的天然铝矾土粉末，主要是由三水铝矿、勃姆石和高岭石等单矿物颗粒组成。高岭石含量不超过总量的25%。分离的粉末在水的存在下制成小球。除水后得到颗粒产品。

18.4.4 轻质陶粒

轻质支撑剂的密度小于2.60g/cm³。是由高岭土和轻质料烧制而成的。需要特殊的焙烧特殊条件（Lemieux和Rumpf，1990，1993，1994）。氧化铝的含量在25%和40%之间（Sweet，1993）。目前已经有密度小于1.3g/cm³的高强度支撑剂的报道（Bienvenu，1996）。

超轻陶粒使常规压裂液在最小黏度时具有优异的输送能力，以确保所需的有效支撑裂缝导流能力（Cawiezel和Gupta，2010）。这些超轻陶粒在黏弹性泡沫流体中的使用，为水力压裂提供了最佳的支撑剂铺置和优良的返排能力。当然，这些液体必须经过优化。这些复合系统需要实验室测试进行表征和优化，以满足超低渗透率油藏的要求。

还可以用硅线石制备支撑剂（Windebank等，2010）。硅线石矿物可以选自蓝晶石、硅线石和红柱石系列。

18.4.4.1 计算和实验方法

轻质陶粒的计算和实验评估已确定其是有效的，并且可以在油气井水力压裂作业中代替石英砂（Kulkarni和Ochoa，2012）。

花生壳、铝或陶瓷颗粒的混合物能降低压裂液的黏度，提高其抗压缩能力。已经用动态有限元分析方法研究了支撑带的准静态压缩，其中每个颗粒都作为独立的模型。

已经研究了各种软、硬颗粒混合物与形状、粒径和颗粒间内摩擦之间的函数关系。粒子间的相互作用表明，孔隙空间的变化是压力、混合物组成和摩擦力的函数。

通过限制粒子的重排获得了更高孔隙度的结果。该模型表明，较软的岩石与硬和软颗粒可以抑制返吐，但同时也会降低支撑带的渗透率（Kulkarni和Ochoa，2012）。

18.4.4.2 反向支撑剂对流

在成熟油田中，面临着最大油气采收率与量小含水的挑战。水的生产会导致几个问题，包括结垢、微粒运移或砂埋、管道腐蚀和静载荷增加（dos Santos等，2009）。

出水几乎是石油生产的必然结果，通常是尽可能推迟出水或含水上升。适当的增产措施对于大多数商业油藏是必要的，包括水驱油藏或近水区域的泥质砂岩和低渗透储层。成熟油田中，压裂液中可以使用相对渗透率改进剂，一次压裂作业同时提高产量并降低含水，达到避开水的目的。

不过，在底水油藏中，人工裂缝可能会形成产水通道，造成水淹。例如，支撑剂的对流和沉降可导致高砂比阶段，从射孔孔眼迅速向下移动到裂缝底部。这可能发生在需要的前置液量较大，支撑剂浓度高或者密度差变化阶段。为了避免这一问题，反向支撑剂对流是一重要技术。需要支撑剂在选定的压裂液中可以漂浮,使用密度从 $1.054g/cm^3$ 到 $1.75g/cm^3$ 的超轻陶粒使之成为可能。

该技术要求泵注的前置液密度高于携砂液的密度,而携砂液的密度略高于超轻支撑剂的密度(dos Santos 等,2009)。

18.4.5 纤维多孔带

可以用纤维和支撑剂的混合物在地层中建立多孔带。纤维材料可以是任何合适的材料,如天然的或合成的有机纤维、玻璃纤维、陶瓷纤维、碳纤维。

多孔带可以过滤掉不需要的颗粒、支撑剂和粉末,同时仍然保持原油正常生产。用纤维建立多孔带以及在地层中铺置支撑剂可以降低设备能耗。泵注纤维和支撑剂可以显著降低限制泵注含砂流体时的泵送阻力(Card 等,2001)。

18.4.6 覆膜支撑剂

通常情况下,悬浮在压裂液中的分级支撑剂等颗粒物在压裂液变稀并返回排至地面时会铺置在裂缝中。这些固体或支撑剂颗粒,可以防止裂缝完全闭合,以形成导流通道,确保碳氢化合物可流出生产(Dusterhoft 等,2008)。

为了防止支撑剂颗粒和其他颗粒随着产出液而回流,支撑剂可以涂敷可固化树脂或可以促使裂缝中支撑剂颗粒固化的增黏剂。部分闭合的裂缝应力可以促使涂层支撑剂的颗粒相互接触,而树脂和增黏剂可以提高支撑剂颗粒之间的连接强度。

具有增黏剂的压力作用保证了支撑剂颗粒固结成具有抗压和抗拉强度的渗透体,同时允许支撑剂带表面有少量的变形,以降低局部负荷的影响或降低支撑剂破碎率(Dusterhoft 等,2008)。

环氧树脂的组分通常包括低聚双酚 A 环氧氯丙烷、4,4′-二氨基二苯基砜固化剂、溶剂、硅烷偶联剂和表面活性剂(Nguyen 等,2007)。

在一系列实验中,用了三种不同的支撑剂颗粒和两组高温环氧树脂进行了评估(Dusterhoft 等,2008)。在每个实验中,树脂的用量均为3%。支撑剂颗粒的类型包括铝土矿、中等强度支撑剂和超轻支撑剂。这些支撑剂耐压 40~80MPa。所有测试的试验温度为120℃。

在数天的周期内应力从 14MPa 不断增加到 80MPa。同样进行了无覆膜支撑剂颗粒的测试。用 API 线性导流池进行了闭合应力和流速对树脂处理后支撑剂性能影响的试验。每个支撑剂充填带的导流能力和渗透率都在14MPa(2000psi)和120℃(250°F)下,连续监测至少 25~30h。

所有的三个支撑剂,裂缝的导流能力和支撑剂充填层的渗透率比未覆膜支撑剂有显著增加。裂缝导流能力和支撑剂充填层渗透率的改善在低应力条件下是明显的。有证据显示,覆膜支撑剂在支撑剂和地层之可以形成更稳定的界面。

可以在支撑剂颗粒上单独涂覆热固性涂料。该涂层可以提高支撑剂的耐化学性。

如果支撑剂对压裂液添加剂不稳定,改进是必要的,如酸类破胶剂。建议在氧化破胶剂存在条件下,可以用可溶性酚醛树脂做涂层材料(Dewprashad,1995)。表 18.2 中列出了可用于支撑剂的聚合物涂层材料。

表 18.2 支撑剂的高分子涂层

材料	参考文献
酚醛/呋喃或呋喃树脂①	Armbruster(1987)
酚醛环氧树脂①	Gibb 等(1989)
热解碳涂层①	Hudson 和 Martin(1989)
双酚树脂②	
酚醛树脂	Johnson 等(1993)
呋喃甲醇树脂②	Ellis 和 Surles(1997)
双酚 A 树脂(可固化的)②	Johnson 和 Tse(1996)
$N-\beta-$(氨基乙基)$-\delta-$氨基丙基三甲氧基硅烷交联环氧树脂	Nguyen 等(2001)
聚酰胺及其他	Nguyen 和 Weaver(2001)

① 耐化学性。
② 预防回流。

覆膜的另一优势是可以减少支撑剂颗粒的摩擦。在这种情况下,具有低摩擦系数的涂层材料为涂层的混合物(de Grood 和 Baycroft,2010)。表 18.3 汇总了这些材料。当然,并不是表 18.3 中列出的所有材料是经济的。多涂层的最终产品的表面是光滑、均匀的。

表 18.3 降阻材料(de Grood 和 Baycroft,2010)

材料	材料
三氧化锑	铋
硼酸	氟化钡钙
铜	石墨
铟	氧化铅
硫化铅	二硫化钼
铌联硒化物	含氟聚合物
聚四氟乙烯	银
锡	二硫化钨

18.4.7 防沉降添加剂

当产生人工裂缝时,水力压裂中支撑剂的传输有两部分。水平分量表现为流速和相关的流线,其可以携带支撑剂至裂缝端部。垂直分量是颗粒沉降速度,是支撑剂粒径和密度以及流体黏度和密度的函数(Watters 等,2010)。

通过注入比流体密度低的添加剂,可以在裂缝中形成多种密度梯度。向上移动的低密度添加剂会干扰高密度支撑剂的向下运动,反之亦然。

支撑剂和添加剂在受限狭缝中的相互干扰会显著阻碍高密度支撑剂的沉降。利用这一原理,可以控制支撑剂的沉降时间。

低密度材料应具有类似于标准支撑剂的粒径分布。除了可以上浮外,这种材料也可以作为支撑剂。这种材料的例子是聚乳酸颗粒或玻璃珠。表 18.4 中给出了这种低密度添加剂对沉降时间的影响。

表 18.4 低密度添加剂中的沉降时间(Watters 等,2010)

距离(ft)	沉降时间(s)	
	(0%)	(5%)
1	8	11
2	22	42
3	47	61
4	56	77

试验是用 1.8g/m³ 的线性胶瓜尔胶中进行的。凝胶的表观黏度在 $511s^{-1}$ 下为 $8mPa·s$。测试的支撑剂浓度为 $240kg/m^3$ 粒径为 30~50 目。

18.4.8 支撑剂回流

压裂施工后支撑剂的回流是个大问题,因为会损坏设备和影响生产。已经有文献讨论了支撑剂回流的机理和控制方法(Nguyen 等,1996b)。为了减少支撑剂回流,可以应用可固化树脂涂层支撑剂(Nimerick 等,1990)。必须把这种支撑剂铺置在生产层位以防止或至少减少支撑剂的回流。

18.4.8.1 热塑性薄膜

已经开发出热塑性薄膜材料以减少压后支撑剂的回流(Nguyen 等,1996a,1996b)。切成薄片的热收缩膜可以减少回流,适用于较宽的温度、闭合应力范围,而对裂缝导流能力影响较小,但有赖于浓度、温度和闭合应力。

18.4.8.2 胶黏包覆材料

添加支撑剂胶黏包覆材料来可以降低颗粒物的回流(Caveny 等,1996)。这种胶黏包覆材料可以是无机纤维或有机纤维。支撑剂颗粒胶黏包覆材料之间的相互机械作用可以防止颗粒回流至井筒。

还可以用聚氨酯涂层达到支撑剂固化,在压后,会由于加聚作用而慢慢聚合(Wiser-Halladay,1990)。

18.4.8.3 可吸附材料

珠子、纤维、条带或颗粒等可吸附材料,可以与支撑剂一起铺置。可吸附材料运移至支撑带的孔道中并聚集成簇,在孔道中吸附结合在一起,这反过来又在地层中促进渗透性砂桥的形成。

可吸附材料与支撑剂砂桥可以减缓并最终防止支撑剂及地层沙的回流,同时仍允许油气通过裂缝保持足够高的产量(Clark 等,2000)。类似的,在支撑剂中铺置纤维束也可以作为压裂防回流材料(Nguyen 和 Schreiner,1999)。

文献中的商品名称

商品名	描述	供应商
SandWedge® NT	基于 2-甲氧基甲基乙氧基丙醇的黏性化合物(Nguyen 等,2007)	Halliburton Energy Services

参 考 文 献

Andrews, W. H., 1987. Bauxite proppant. US Patent 4 713 203, assigned to Comalco AluminiumLtd., Victoria, AU, 15 December 1987. < http://www.freepatentsonline.com/4713203.html >.

Andrews, W. H., 1988. Sintered bauxite pellets and their application as proppants in hydraulicfracturing. AU Patent 579 242, 17 November 1988.

Armbruster, D. R., 1987. Precured coated particulate material. US Patent 4 694 905, 22 September1987.

Asadi, M., Woodroof, R. A., Himes, R. E., 2008. Comparative study of flowback analysis usingpolymer concentrations and fracturing – fluid tracer methods: a field study. SPE Prod. Oper. 23(2), 147 – 157. http://dx.doi.org/10.2118/101614 – PA.

Bienvenu Jr., R. L., 1996. Lightweight proppants and their use in hydraulic fracturing. US Patent5 531 274, 2 July 1996.

Card, R. J., Howard, P. R., Feraud, J. P., Constien, V. G., 2001. Control of particulate flowback insubterranean wells. US Patent 6 172 011, assigned to Schlumberger Technol. Corp., 9 January 2001.

Caveny, W. J., Weaver, J. D., Nguyen, P. D., 1996. Control of particulate flowback in subterranean wells. US Patent 5 582 249, 10 December 1996.

Cawiezel, K. E., Gupta, D. V. S., 2010. Successful optimization of viscoelastic foamed fracturingfluids with ultralight-weight proppants for ultralow – permeability reservoirs. SPE Prod. Oper. 25(1), 80 – 88. http://dx.doi.org/10.2118/119626 – PA.

Clark, M. D., Walker, P. L., Schreiner, K. L., Nguyen, P. D., 2000. Methods of preventingwell fractureproppant flowback. US Patent 6 116 342, assigned toHalliburton Energy Service, 12 September 2000.

de Grood, R. J. C., Baycroft, P. D., 2010. Use of coated proppant to minimize abrasive erosion in highrate fracturing operations. US Patent 7 730 948, assigned to Baker Hughes Incorp., Houston, TX, 8 June 2010. < http://www.freepatentsonline.com/7730948.html >.

Dewprashad, B., 1995. Method of producing coated proppants compatible withoxidizing gelbreakers. US Patent 5 420 174, assigned to Halliburton Co., 30 May 1995.

dos Santos, J. A. C. M., Cunha, R. A., de Melo, R. C. B., Aboud, R. S., Pedrosa, H. A., Marchi, F. A., 2009. Inverted – convection proppant transport for effective conformance fracturing. SPE Prod. Oper. 24(1), 187 – 193. http://dx.doi.org/10.2118/109585 – PA.

Duenckel, R., Conway, M. W., Eldred, B., Vincent, M. C., 2012. Proppant diagenesis – integratedanalyses provide new insights into origin, occurrence, and implications for proppant performance. SPE Prod. Oper. 27(2), 131 – 144.

Dusterhoft, R. G., Fitzpatrick, H. J., Adams, D., Glover, W. F., Nguyen, P. D., 2008. Methods ofstabilizing surfaces of subterranean formations. US Patent 7 343 973, assigned to Halliburton

Energy Services, Inc., Duncan, OK, 18 March 2008. < http://www.freepatentsonline.com/7343973.html >.

Ellis, P. D., Surles, B. W., 1997. Chemically inert resin coated proppant system for control of proppant flowback in hydraulically fractured wells. US Patent 5 604 184, assigned to Texaco Inc., 18February 1997.

Fitzgibbon, J. J., 1986. Use of uncalcined/partially calcined ingredients in the manufacture ofsintered pellets useful for gas and oil well proppants. US Patent 4 623 630, 18 November1986.

Fitzgibbon, J. J., 1988. Sintered, spherical, composite pellets prepared from clay as a majoringredient useful for oil and gas well proppants. CA Patent 1 232 751, 16 February 1988.

Fitzgibbon, J. J., 1989. Sintered spherical pellets containing clay as a major component useful forgas and oil well proppants. US Patent 4 879 181, 7 November 1989.

Gibb, J. L., Laird, J. A., Berntson, L. G., 1989. Novolac coated ceramic particulate. EPPatent 308 257, 22 March 1989.

Gibb, J. L., Laird, J. A., Lee, G. W., Whitcomb, W. C., 1990. Particulate ceramic useful as a proppant. US Patent 4

944 905,31 July 1990.

Hudson, T. E. , Martin, J. W. ,1989. Pyrolytic carbon coating of media improves gravel packing andfracturing capabilities. US Patent 4 796 701,10 January 1989.

Johnson, C. K. , Tse, K. T. ,1996. Bisphenol – containing resin coating articles and methods of usingsame. EP Patent 735 234, assigned to Borden Inc. ,2 October 1996.

Johnson, C. R. , Tse, K. T. , Korpics, C. J. ,1993. Phenolic resin coated proppants with reducedhydraulic fluid interaction. US Patent 5 218 038, assigned to Borden, Inc. , Columbus, OH,8 June 1993. < http://www. freepatentsonline. com/5218038. html >.

Khaund, A. ,1987a. Sintered low density gas and oil well proppants from a low cost unblendedclay material of selected composition. US Patent 4 668 645,26 May 1987.

Khaund, A. K. , 1987b. Stress – corrosion resistant proppant for oil and gas wells. US Patent 4 639427, 27 January 1987.

Kulkarni, M. C. , Ochoa, O. O. ,2012. Mechanics of light weight proppants: a discrete approach. Compos. Sci. Technol. 72(8),879 – 885. http://dx. doi. org/10. 1016/j. compscitech. 2012. 02. 017.

Laird, J. A. , Beck, W. R. ,1989. Ceramic spheroids having low density and high crush resistance. EP Patent 207 668, 5 April 1989.

Lemieux, P. R. , Rumpf, D. S. ,1990. Low density proppant and methods for making and using same. EP Patent 353 740,7 February 1990.

Lemieux, P. R. , Rumpf, D. S. , 1993. Lightweight oil and gas well proppant and methods for makingand using same. AU Patent 637 576,3 June 1993.

Lemieux, P. R. , Rumpf, D. S. , 1994. Lightweight proppants for oil and gas wells and methods formaking and using same. CA Patent 1 330 255,21 June 1994.

Nguyen, P. D. , Weaver, J. D. , Parker, M. A. , King, D. G. ,1996a. Thermoplastic film prevents proppant flowback. Oil Gas J. 94(6),60 – 62.

Nguyen, P. D. , Weaver, J. D. , Parker, M. A. , King, D. G. , Gillstrom, R. L. , Van Batenburg, D. W. ,1996b. Proppant flowback control additives. In: Proceedings Volume. Annual SPE TechnicalConference, Denver,6 – 9 October 1996, pp. 119 – 131.

Nguyen, P. D. , Schreiner, K. L. ,1999. Preventing well fracture proppant flow – back. USPatent5 908 073, assigned to Halliburton Energy Serv. ,1 June 1999.

Nguyen, P. D. , Weaver, J. D. ,2001. Method of controlling particulate flowback in subterranean wellsand introducing treatment chemicals. US Patent 6 209 643, assigned to Halliburton EnergyServ. ,3 April 2001.

Nguyen, P. D. , Weaver, J. D. , Brumley, J. L. ,2001. Stimulating fluid production from unconsolidated formations. US Patent 6 257 335, assigned to Halliburton Energy Serv. ,10 July 2001.

Nguyen, P. D. , Barton, J. A. , Isenberg, O. M. ,2007. Methods and compositions for consolidating proppant in fractures. US Patent 7 264 052, assigned to Halliburton Energy Services, Inc. , Duncan, OK,4 September 2007. < http://www. freepatentsonline. com/7264052. html >.

Nimerick, K. H. , McConnell, S. B. , Samuelson, M. L. ,1990. Compatibility of resin – coated proppantswith crosslinked fracturing fluids. In: Proceedings Volume. 65th Annual SPE Technical Conference, New Orleans,23 – 26 September 1990, pp. 245 – 250.

Sullivan, R. , Woodroof, R. , Steinberger – Glaser, A. , Fielder, R. , Asadi, M. ,2004. Optimizingfracturing fluid cleanup in the Bossier sand using chemical frac tracers and aggressive gelbreaker deployment. In: Proceedings of SPE Annual Technical Conference and Exhibition. Society of Petroleum Engineers. http://dx. doi. org/10. 2118/90030 – MS.

Sweet, L., 1993. Method of fracturing a subterranean formation with a lightweight propping agent. US Patent 5 188 175, 23 February 1993.

Todd, B. L., Slabaugh, B. F., Munoz Jr., T., Parker, M. A., 2006. Fluid loss control additives foruse in fracturing subterranean formations. US Patent 7 096 947, assigned to HalliburtonEnergy Services, Inc., Duncan, OK, 29 August 2006. <http://www.freepatentsonline.com/7096947.html>.

Watters, J. T., Ammachathram, M., Watters, L. T., 2010. Method to enhance proppant conductivityfrom hydraulically fractured wells. US Patent 7 708 069, assigned to Superior Energy Services, L. L. C., New Orleans, LA, 4 May 2010. <http://www.freepatentsonline.com/7708069.html>.

Windebank, M., Hart, J., Alary, J. A., 2010. Proppants and anti-flowback additives madefrom sillimanite minerals, methods of manufacture, and methods of use. US Patent7 790 656, assigned to Imerys, Paris, FR, 7 September 2010. <http://www.freepatentsonline.com/7790656.html>.

Wiser-Halladay, R., 1990. Polyurethane quasi prepolymer for proppant consolidation. US Patent4 920 192, 24 April 1990.

19 特殊添加剂

下面介绍一些性能卓越的压裂液添加剂。

19.1 自生热系统

压裂处理低温、浅层和高凝油油藏时,主要是解决破胶不彻底、压裂液无法完全返排以及注入冷流体造成的地层冷伤害的问题。

为了避免这些问题,研发了封装的自生热水力压裂液体系。油气生产中常用两种自生热系统。一种是过氧化氢系统,反应如下(Bayles1s,2000):

$$H_2O_2 \longrightarrow \frac{1}{2}O_2 + H_2O \tag{19.1}$$

该系统的标准反应热,ΔH 为 $-196kJ/mol$。亚硝酸铵盐自生热系统的反应如下(Wu 等,2005):

$$NO_2^- + NH_4^+ \longrightarrow N_2 + H_2O \tag{19.2}$$

该系统的标准反应热,ΔH 为 $-333kJ/mol$,远高于第一个系统。

深入研究了第二个系统,草酸作为催化剂,并用相分离的方法将乙基纤维素和石蜡作为涂层材料。该配方中加入环己烷,连续搅拌并加热至81℃,直至乙基纤维素溶解。搅拌30min,冷却,中间产物过滤回收,用环己烷清洗,然后干燥。

该系统的反应动力学可以描述为:

$$\frac{dc}{dt} = -1.267 \times 10^7 C_H^{1.17} C_0^{1.88} \exp\left(-\frac{5360}{T}\right) \tag{19.3}$$

式中,$\frac{dc}{dt}$ 是反应物的消耗率,$mol/(L \cdot min)$;C_H 是质子催化剂的浓度,mol/L;C_0 为初始浓度,mol/L。表达式类似于阿伦尼乌斯因子。

表 19.1 给出了压裂液配方。结果表明,含有封装发热剂的水力压裂液具有良好的稳定性和配伍性。

表 19.1 压裂液配方(Wu 等,2005)

配方	浓度(%)
羟丙基瓜尔胶	0.6
硼砂	0.7
过硫酸铵	0.08

续表

生热剂	浓度(mol/L)
试验1	1.5
试验2	1.75
试验3	2.0

当压裂液中含有 2.0mol/L 发热剂，其中包裹的内含物为 0.93% 草酸和 0.08% 为过硫酸铵时，最高温度可达 78℃，4h 后残留液的黏度为 3.12mPa·s(Wu 等,2005)。

19.2 可交联的合成聚合物

目前已经研发出使用温度高达 232℃ 的水力压裂液(Holtsclaw 和 Funkhouser,2010)。这种流体技术使用合成聚合物，该聚合物可以用金属离子交联形成高黏度液体。该聚合物压裂冻胶克服了传统瓜尔胶及其衍生物压裂液的温度限制。一般使用的是基于丙烯酸、丙烯酰胺和 2-丙烯酰氨基-2-甲基-1-丙磺酸的三元共聚物。

研发了在高温下具有良好稳定性的压裂液，可以在最苛刻的环境中保证支撑剂更好地传输和铺置。交联反应可以在 38~138℃ 之间进行调节。从而允许根据特殊井况的优化。

用流变学数据表征流体的稳定性、交联性能和控制流体破胶。此外，动态流体滤失和导流能力恢复数据说明了支撑剂带中流体的返排性能(Holtsclaw 和 Funkhouser,2010)。

19.3 单相微乳液

已经研发出单相微乳液和胶凝聚合物相结合的压裂液系统(Liu 等,2010)。此配方适用于降低聚合物用量。并且由于协同作用而改善聚合物凝胶的性能。

从相应的相图数据可以得到单相微乳液体系的配方。该配方是在各种浓度的胶凝聚合物体系中加入单相微乳液，从而使制得的液体具有高黏度、低滤失量和低摩阻的特点。

此外，该凝胶体系在地层中具有低残渣、低表面张力、低返排启动压力和高岩心渗透率保持率的特点。该配方有望在压裂中减少地层伤害、降低返排启动压力，并保持原有的岩心恢复。

19.4 复合交联剂

可以用基于锆的三乙醇胺络合物、四三乙醇胺锆和四(烷基)乙二胺组成复合交联剂(Putzig,2010a)。

可以用四丙基锆在乙醇溶液中与三乙醇胺反应，合成四三乙醇胺锆。然后，加入乳酸，得到锆醇胺羧酸络合物。已经有详细的制备方法描述(Putzig,2010b)。

参 考 文 献

Bayless,J. H. ,2000. Hydrogen peroxide applications for the oil industry. World Oil 221(5) ,50-54.
Holtsclaw,J. , Funkhouser,G. P. ,2010. A crosslinkable synthetic – polymer system for hightemperature hydraulic –

fracturing applications. SPE Drill. Completion 25 (4), 555 – 563. <http://www.onepetro.org/mslib/servlet/onepetropreview? id = SPE – 125250 – PA>.

Liu, D. – X., Fan, M. – F., Yao, L. – T., Zhao, X. – T., Wang, Y. – L., 2010. A new fracturing fluid with combination of single phase microemulsion and gelable polymer system. J. Petrol. Sci. Eng. 73 (3 – 4), 267 – 271. http://dx.doi.org/10.1016/j.petrol.2010.07.008.

Putzig, D. E., 2010a. Hydraulic fracturing methods using cross – linking composition comprising zirconium triethanolamine complex. US Patent 7 730 952, assigned to E. I. duPont de Nemours and Co., Wilmington, DE, 8 June 2010. <http://www.freepatentsonline.com/7730952.html>.

Putzig, D. E., 2010b. Process to prepare zirconium – based cross – linker compositions and their use in oil field applications. US Patent 7 754 660, assigned to E. I. duPont de Nemours and Co., Wilmington, DE, 13 July 2010. <http://www.freepatentsonline.com/7754660.html>.

Wu, J., Zhang, N., Wu, X., Liu, X., 2005. Experimental research on a new encapsulated heat – generating hydraulic fracturing fluid system. Chin. J. Geochem. 25 (2), 162 – 166. http://dx.doi.org/10.1007/BF02872176.

20 环境因素

20.1 风险分析

近年来,通过水平钻井和水力压裂技术的应用,使得页岩气地层开发在经济上可行。由于这些技术使用了大量的水,从而存在着潜在的环境风险,也因此存在水资源污染的重大风险。

使用概率方法对位于美国东北部的伊利湖和安大略湖南部的马塞勒斯页岩中天然气开采过程中潜在的水污染进行了评估(Rozell 和 Reaven,2012)。

当数据零散且参数高度不确定时,使用概率边际分析方法比较适合。可以确定水污染的五条途径(Rozell 和 Reaven,2012):

(1)运输溢洒;
(2)井筒泄漏;
(3)通过破碎岩石渗漏;
(4)井场排放;
(5)废水处理。

生成了每种途径的概率框图。与水力压裂废水处理有关的潜在的污染风险,以及认识的不确定性风险,比其他途径形成的风险大几个数量级。

即使在最好的情况下,单口井也很可能排放至少 $200m^3$ 受污染的液体。马塞勒斯页岩区的井的总数可达数万口,有了这种潜在的实际风险,必须要求采取额外措施以便减小受污染流体泄漏的可能性。为了减少大量的主观不确定性因素,需要收集更多的从用过的压裂液中消除污染物的工业和城市污水处理设施方面的相关资料(Rozell 和 Reaven,2012)。

应用质量平衡原则,处理用于马塞勒斯页岩的 47 口天然气井的压裂液添加剂中的 12 种不同有害成分。假设 1000gal 稀释添加剂排放到地表水体或土壤中,可以对溢散情况进行建模。就所得到的浓度按照大小排序,对每个分区的数量进行比较。氢氧化钠、4,4-二甲基噁唑烷和盐酸存在于水、土壤以及生物中的质量浓度最高。

4,4-二甲基噁唑烷在空气中含量最高(Aminto 和 Olson,2012)。

20.2 污水回收

已经开发了与产出水有关的污水回收系统和方法,以便生产质量达标并且能够重新用作压裂水的净化水(Shafer 等,2009 年)。

产出水含有来自于地下环境的天然污染物,例如来自于含油或含气地层的烃类和无机盐。产出水还可能含有人造污染物,其中可能有包含聚合物和无机交联剂、聚合物破胶剂、降阻剂和人工润滑剂的废弃压裂液。当产出水又被甲醇污染后,将存在一个特殊问题,其几乎可以穿透任何可用的膜过滤系统。

污染水回收系统包括先对污染水进行厌氧消化,然后充入气体以提高生物消化能力。充气后,采用浮选作业,有效去除余下的降阻剂,使净化水得到回收并重新用作压裂配液用水(Shafer 等,2009 年)。在独立的设备中,经过一个砂粒充填过滤器,然后进入一系列生物反应器,最后进入硼砂处理装置。经过这样处理后,清洁的废物可以排放到环境中。

表 20.1 中给出了从这样一个系统中获得的污染物浓度的典型值和目标值。

表 20.1　污染物的浓度(Shafer 等,2009)

污染物	典型值	目标(最大值)
总溶解固体量(mg/mL)180℃	9000~16000	10000
总溶解固体量(mg/mL)105℃	0~100	75
总有机碳(mg/mL)	400~800	700
化学需氧量(mg/mL)	1000~3000	2000
生物需氧量(mg/mL)	500~1500	1000
铁	1~10	5
氯化物	5000~10000	6000
钾	100~500	300
钙	50~250	150
镁	10~100	25
钠	2000~5000	3000
硫酸盐	40~200	50
碳酸盐	0~100	25
重碳酸盐	100~1200	800
硼	0~20	15

20.2.1　水力压裂产生的废水

水力压裂可能产生大量的废水。其工艺涉及到注入水、压裂砂和化学品的混合物。该工艺需要将压力超过 20MPa,流速超过 5L/s 的水泵入裂缝,以便形成长裂缝砂粒充填,与页岩中的天然裂缝沟通。总体而言,每口井水力压裂工艺需要使用 10000m^3 的水(DiTommaso 和 DiTommaso,2012)。

以下描述了一种用于净化处理钻井完井之后,来自于水力压裂的返排液的方法。该方法包括(DiTommaso 和 DiTommaso,2012):

(1)返排液中除油;

(2)使用一个孔径大约为 0.1μm 或更小的超滤器来过滤返排液,以便除去水中固的体颗粒和有机大分子,如苯、乙苯、甲苯和二甲苯;

(3)浓缩返排液并产生一种浓盐水,其中含盐量为浓盐水总重的 15%~40%;

(4)使用有效量的药剂进行一次或多次化学沉淀处理,以便沉积出所需的高品质的商用产品,如硫酸钡、碳酸锶或碳酸钙;

(5)对经过化学处理并浓缩的返排浓盐水进行结晶处理,以生产出纯的盐产品,如氯化钠和氯化钙。

20.2.2 从返排液中回收磷

在压裂过程中,使用磷酸酯作为胶凝剂。在这些操作过程中,磷可以溶于原油并随之产出。然而,磷酸酯胶凝剂的挥发性磷成分,会引起炼油装置堵塞,而这是一种潜在的危险。

建议将原油中最大挥发性磷的含量控制在 0.5mg/L,以避免意外致使炼油设备停机,并付出高昂的代价。该标准通过综合应用新的化学反应和一般现场稀释作业,已经加入到原油中的胶凝所含磷的平均浓度,并假定原来的石油不含磷确定的。

在对返排流体的研究中,观察到总的和挥发性的磷浓度远远大于添加的磷浓度。已有研究给出了磷浓度显著高于原添加磷浓度的可能原因(Fyten 等,2007)。

已经开发了用于原油馏分中总挥发性磷的分析,该分析基于感应耦合等离子体发射光谱方法(Nizio 和 Harynuk,2012)。然而,该方法受到精度差和最高检测浓度上限 0.5mg/mL 的制约。

另一种方法是用带有氮磷检测器的全二维气相色谱仪进行检测。此方法为工业石油样品中的磷酸烷基酯检测,提供了高精度的定性和定量分析。介绍了压裂液样品回收研究以及四种回收压裂液或返排原油混合中的磷酸烷基酯的研究概要(Nizio 和 Harynuk,2012)。

20.3 绿色配方

20.3.1 生物可降解螯合剂

当今石油公司要求应用"绿色"压裂液化学进行处理。常用的破乳剂和阻垢剂往往有环境问题,且不具有多功能特性。

生物可降解和无毒的螯合剂可以在水基压裂液中,通过离子的螯合完成多种有益的功能。包括各种组合的多种功能如下(Crews,2006):

(1)破乳剂;
(2)破乳增强剂;
(3)阻垢剂;
(4)交联延迟剂;
(5)交联凝胶稳定剂;
(6)酶破胶剂;
(7)稳定剂。

20.3.2 无毒返排配方

当注入裂缝性储层时,无毒、环保、绿色返排助剂可减少水的堵塞(Berger 和 Berger,2011)。

制备 20.1:返排助剂是将 15 份乳酸乙酯溶解到 15 份甲基树脂酸盐中制备得到的。用含有 6mol 环氧乙烷的 12.5 份月桂醇与 12.5 份 70% 的十二基磺基丁二酸钠的水溶液混合成第二种溶液。最后,将这两种溶液在搅拌下混合,并缓慢地加入 55 份水,以便形成透明低黏度的稳定溶液。

20.3.3 交联剂

在考虑到运输及环保方面,ⅣB 族金属醇盐交联剂既不易燃又环保(Vaughn 等,2011)。表 20.2 中列出了一些可商业化的产品。

表20.2 环保交联剂(沃恩等,2011)

化学描述	商品名
C_6醇烷氧基锆	Vertec™ XL985
四(2-乙基己基)锆酸酯	
C_4乙二醇烷氧基锆	Vertec™ XL980
C_6醇烷氧基钛	Vertec™ XL121
四(2-乙基己基)钛酸酯	
C_4乙二醇钛醇盐	Vertec™ XL990

已经有报告详细描述了由这些化合物制备交联剂的几种方法及其流变学性能和阻燃程度(Vaughn等,2011)。

20.4 自降解泡沫成分

描述了依赖于pH值的泡沫压裂液。这种压裂液是由胶凝剂、表面活性剂和支撑剂混合配制而成的。表面活性剂能够促进压裂液在初始pH值环境中发泡,并在pH值发生改变时消泡。压裂液的pH值在就地条件下与酸性物质接触会发生变化,导致压裂液中的泡沫质量降低。随着泡沫质量降低,压裂液支撑剂沉积到岩层中形成的裂缝内(Chatterji等,2005)。可以将压裂液返排回地面回收。通过恢复其pH值,压裂液再发泡,并再次注入井下。

表面活性剂既可以是两性的,也可以是阴离子的。合适的表面活性剂是叔烷基胺乙氧基化物。在发泡表面活性剂中加入氢离子,即可以将发泡表面活性剂转变为消泡表面活性剂。加入氢氧化物,则又逆转为发泡剂。质子化反应如图20.1所示。

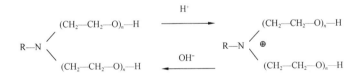

图20.1 质子化反应(Chatterji等,2005)

自降解泡沫是一种由阴离子表面活性剂和非离子表面活性剂组成的混合物。在泡沫压裂液回收过程中,形成本质上不太稳定的泡沫(Dahanayake等,2008)。

在自降解发泡成分中,优选的阴离子表面活性剂是月桂酰胺单乙醇胺磺基丁二酸二钠。非离子表面活性剂,可以用聚乙二醇羟乙基化脂肪酸酯。已证实,在pH值高的情况下,泡沫劣质化速度加快。

文献中的商品名称

商品名	描述	供应商
Lewatit®(系列)	二乙烯基苯/苯乙烯共聚物离子交换树脂(Shafer等,2009)	Lanxess Deutschland GmbH
Vertec™(系列)	环保交联剂(Vaughn等,2011)	

参 考 文 献

Aminto, A. , Olson, M. S. , 2012. Four – compartment partition model of hazardous componentsin hydraulic fracturing fluid additives. J. Nat. Gas Sci. Eng. 7, 16 – 21. http://dx. doi. org/10. 1016/j. jngse. 2012. 03. 006.

Berger, P. D. , Berger, C. H. , 2011. Environmental friendly fracturing and stimulation composition and method of using the same. US Patent 7 998 911, assigned to Oil Chem. Tech. , Sugar Land, TX, 16 August 2011. < http://www. freepatentsonline. com/7998911. html > .

Chatterji, J. , King, K. L. , King, B. J. , Slabaugh, B. F. , 2005. Methods of fracturing a subterranean formation using a pH dependent foamed fracturing fluid. US Patent 6 966 379, assignedto Halliburton Energy Services, Inc. , Duncan, OK, 22 November 2005. < http://www. freepatentsonline. com/6966379. html > .

Crews, J. B. , 2006. Biodegradable chelant compositions for fracturing fluid. US Patent 7 078 370, assigned to Baker Hughes Incorp. , Houston, TX, 18 July 2006. < http://www. freepatentsonline. com/7078370. html > .

Dahanayake, M. S. , Kesavan, S. , Colaco, A. , 2008. Method of recycling fracturing fluids using aself – degrading foaming composition. US Patent 7 404 442, assigned to Rhodia Inc. (Cranbury, NJ), 29 July 2008. < http://www. freepatentsonline. com/7404442. html >

DiTommaso, F. A. , DiTommaso, P. N. , 2012. Method of making pure salt from frac – water/wastewater. US Patent 8 273 320, assigned to FracPure Holdings LLC, Dover, DE, 25September 2012. < http://www. freepatentsonline. com/8273320. html > .

Fyten, G. , Houle, P. , Taylor, R. S. , Stemler, P. S. , Lemieux, A. , 2007. Total phosphorus recovery inflowback fluids after gelled hydrocarbon fracturing fluid treatments. J. Can. Petrol. Technol. 46 (12), 17 – 21. http://dx. doi. org/10. 2118/07 – 12 – TN2.

Nizio, K. D. , Harynuk, J. J. , 2012. Analysis of alkyl phosphates in petroleum samples bycomprehensive two – dimensional gas chromatography with nitrogen phosphorus detection andpost – column deans switching. J. Chromatogra. A 1252, 171 – 176. http://dx. doi. org/10. 1016/j. chroma. 2012. 06. 070.

Rozell, D. J. , Reaven, S. J. , 2012. Water pollution risk associated with natural gas extraction fromthe Marcellus Shale. Risk Anal. 32 (8), 1382 – 1393. http://dx. doi. org/10. 1111/j. 1539 – 6924. 2011. 01757. x.

Shafer, L. L. , James, J. W. , Rath, R. D. , Eubank, J. , 2009. Method for generating fracturing water. US Patent 7 527 736, assigned to Anticline Disposal, LLC, Rapid City, SD, 5 May 2009. < http://www. freepatentsonline. com/7527736. html > .

Vaughn, D. E. , Duncan, R. H. , Harry, D. N. , Williams, D. A. , 2011. Non – flammable, non – aqueousgroup ivb metal alkoxide crosslinkers and fracturing fluid compositions incorporating same. USPatent 7 879 771, assigned to Benchmark Performance Group, Inc. , Houston, TX, 1 February2011. < http://www. freepatentsonline. com/7879771. html > .

附　录

商品名称

Aerosil®
气相二氧化硅(Fumed silica)
Aldacide® G
杀菌剂,戊二醛(Biocide, glutaraldehyde)
Alkamuls® SMO
失水山梨醇油酸酯(Sorbitan monooleate)
Alkaquat™ DMB-451
烷基二甲基苄基氯化铵(Dimethyl benzyl alkyl ammonium chloride)
Amadol® 511
妥尔油脂肪酸二乙醇胺(Tall oil fatty acid diethanolamine)
Benol®
石蜡油(White mineral oil)
Britolo® 35 USP
高黏矿物油(High viscosity mineral oil)
CAB-O-SIL™(Series)
气相二氧化硅(Fumed silica)
Captivates® liquid
鱼明胶和阿拉伯树胶胶囊(Fish gelatin and gum acacia encapsulation coating)
Carbolite™
不同粒径的陶粒支撑剂(Sized ceramic proppant)
Carnation®
石蜡油(White mineral oil)
ClearFRAC™
增产工作液(Stimulating Fluid)
Dacron®
聚对苯二甲酸乙二醇酯(Poly(ethylene terephtthalate)
Dequest® 2060
二乙烯三胺五亚甲基膦酸(Diethylene triamine pentamethylene phosphonic acid)
Diamond FRAQ™
VES破胶剂(VES breaker)
DiamondFRAQ™
VES体系(VES System)
Empol™(Series)
低聚油酸(Oligomeric oleic acid)
Envirogem®
非离子表面活性剂(Nonionic surfactants)
Escaid®(Series)
矿物油(Mineral oils)
Flow-Back™
黄胞胶,韦兰胶(Xanthan gum, welan gum)
Fluorinert™(Series)
氟碳化合物(Fluorocarbons)
Flutec™ PP
氟碳化合物(Fluorocarbon)
FR™(Series)
降阻剂(Friction reducer)
Galden™ LS
氟碳化合物(Fluorocarbon)
Geltone®(Series)
有机黏土(Organophilic clay)
Gloria®
高黏矿物油(High viscosity mineral oil)
Hydrobrite® 200
石蜡油(White mineral oil)
Isopar®(Series)
异烷烃溶剂(Isoparaffinic solvent)
Jeffamine® D-230
聚(氧化丙烯)二胺(Poly(oxypropylene)diamine)
Jeffamine® EDR-148

三甘醇二胺(Triethyleneglycol diamine)
Jeffamine® HK-511
聚氧化烯胺(Poly(oxyalkylene) amine)
Jordapon® ACl
椰油基羟丙基磺酸钠表面活性剂(Sodium cocoyl isothionate surfactant)
Jordapon® Cl
椰油酰羟乙磺酸酯铵表面活性剂(Ammonium cocoyl isothionate surfactant)
Kaydol® oil
矿物油(Mineral oil)
Krytox™
氟化油和油脂(Fluorinated oils and greases)
Lewatit®(Series)
二乙烯基苯/苯乙烯共聚物离子交换树脂(Divinylbenzene/styrene copolymer ion exchange resins)
LPA® -210
有机溶剂(Hydrocarbon solvent)
Microsponge™
多孔物质(Porous solid substrate)
Performance® 225N
基础油(Base Oil)
Poly-S. RTM
聚合物胶囊(Polymer encapsulation coating)
Rhodafac® LO-11A
磷酸酯(Phosphate ester)
Rhodafac® LO-11A-LA
磷酸酯(Phosphate ester)
Rhodafac® RS-410
聚(氧基-1,2-亚乙基)十三烷基羟基磷酸(Poly(oxy-1,2-ethandiyl) tridecyl hydroxy phosphate)
Rhodoclean™
非离子表面活性剂(Nonionic surfactant)
SandWedge® NT
基于2-甲氧基甲基乙氧基丙醇的黏性化合物(Tackifying compound, based on 2-methoxymethylethoxy propanol)
Scaletreat® XL14FD

聚(马来酸盐)(poly(maleate))
Shale Guard™ NCL100
页岩防膨剂(Shale anti-swelling agent)
Span® 20
失水山梨醇月桂酸酯(Sorbitan monolaurate)
Span® 40
失水山梨醇单棕榈酸酯(Sorbitan monopalmitate)
Span® 61
失水山梨醇单硬脂酸酯(Sorbitan monostearate)
Span® 65
失水山梨醇三硬脂酸酯(Sorbitan tristearate)
Span® 80
失水山梨醇单油酸酯(Sorbitan monooleate)
Span® 85
失水山梨醇三油酸酯(Sorbitan trioleate)
Surfonic®
乳化剂,乙氧基化C_{12}醇(Emulsifier, ethoxylated C_{12} alcohol)
SurFRAQ™ VES
牛脂酰氨基丙基胺氧化物(Tallow amido propylamine oxide)
Tegopren™(Series)
硅氧烷乳化剂(Siloxane emulsifier)
Tomadol®
脂肪胺(Fatty amines)
Tween® 20
失水山梨醇月桂酸酯(Sorbitan monolaurate)
Tween® 21
失水山梨醇月桂酸酯(Sorbitan monolaurate)
Tween® 40
失水山梨醇单棕酸酯(Sorbitan monopalmitate)
Tween® 60
失水山梨醇单硬脂酸酯(Sorbitan monostearate)
Tween® 61
失水山梨醇单硬脂酸酯(Sorbitan monostearate)

Tween® 65
失水山梨醇三硬脂酸酯(Sorbitan tristearate)
Tween® 81
失水山梨醇油酸酯(Sorbitan monooleate)
Tween® 85
失水山梨醇油酸酯(Sorbitan monooleate)
Tyzor NPZ®
正丙醇锆酸酯(tetra-n-propyl zirconate)
Vertec™ (Series)
环保交联剂(Environmental friendly crosslinking Agents)
VES-STA 1
凝胶稳定剂(Gel stabilizer)
Wellguard™ 7137
卤间化合物凝胶破胶剂(Interhalogen gel breaker)
WG-3L VES-AROMOX® APA-T
黏弹性表面活性剂(Viscoelastic surfactant)
WS-44
乳化剂(Emulsifier)

有机化合物缩写

AA
丙烯酸(Acrylic acid)
AM
丙烯酰胺(Acrylamide)
AMPS
2-丙烯酰氨基-2-甲基-1-丙烷磺酸衍生物(2-Acrylamido-2-methyl-1-propane sulfonic acid)
CMC
羧甲基纤维素(Carboxymethyl cellulose)
EDTA
乙二胺四乙酸(Ethylenediamine tetraacetic acid)
EG
乙二醇(Ethylene glycol)
EO
环氧乙烷(Ethylene oxide)
HEC
羟乙基纤维素(Hydroxyethyl cellulose)
HPC
羟丙基纤维素(Hydroxypropyl cellulose)

IFT
界面张力(Interfacial tension)
MA
顺丁烯二酸酐(Maleic anhydride)
PAA
聚(丙烯酸)(Poly(acrylic acid))
PAM
聚(丙烯酰胺)(Poly(acrylamide))
PEG
聚(乙二醇)(Poly(ethylene glycol))
PEI
聚酰亚胺(Poly(ether imide))
PO
环氧丙烷(Propylene oxide)
PPCA
膦基聚(羧酸)(Phosphino-poly(carboxylic acid))
PVS
聚(乙烯基磺酸)(Poly(vinyl sulfonate))
VES
黏弹性表面活性剂(Viscoelastic surfactant)

化学药品

乙酸(Acetic acid)
丙烯酰乙氧基三甲基氯化铵(Acryloyloxyethyltrimethyl ammonium chloride)
己二酸(Adipic acid)
1-烯丙氧基-2-羟丙基磺酸(1-Allyloxy-2-hydroxypropyl sulfonic acid)
异丙醇铝(Aluminum isopropoxide)
氨基亚氨基二乙酸(Amidoiminodiacetic acid)

$N-\beta-$（氨基乙基）$-\delta-$氨基丙基三甲氧基硅烷（$N-\beta-$(Aminoethyl)$-\delta-$aminopropyltrimethoxysilane）

氨基三亚甲基膦酸（Aminotri(methylenephosphonic acid)）

过硫酸铵（Ammonium persulfate）

2-丙烯酰氨基-2-甲基-1-丙磺酸（2-Acrylamido-2-methyl-1-propane sulfuric acid）

直链淀粉（Amylose）

对叔戊基苯酚（$p-tert-$Amylphenol）

阿拉伯糖醇（Arabitol）

偶氮胂（Arsenazo）

2,2-偶氮（双脒基丙烷）二盐酸盐（2,2'-Azo(bis-amidinopropane) dihydrochloride）

2,2'-偶氮聚（2-脒基丙烷）二盐酸盐（2,2'-Azobis(2-amidinopropane) dihydrochloride）

苯四甲酸（Benzene tetracarboxylic acid）

苯甲酸（Benzoic acid）

苯甲基十六烷基二甲基溴化铵（Benzylcetyldimethyl ammonium bromide）

甜菜碱（Betaine）

双六亚甲基三胺五（亚甲基膦酸）（Bis-hexamethylene triamine pentakis(methylene phosphonic acid)）

$N,N-$二(2-羟乙基)甘氨酸（$N,N-$Bis(2-hydroxyethyl)glycine）

双[四（羟甲基）鏻]硫酸盐（Bis[tetrakis(hydroxymethyl)phosphonium] sulfate）

2-溴-2-硝基-1,3-丙二醇（2-Bromo-2-nitro-1,3-propanediol）

9-溴硬脂酸（9-Bromo stearate）

溴硝醇（Bronopol）

1,2-环氧丁烷（1,2-Butylene oxide）

过氧化钙（Calcium peroxide）

羧甲基羟乙基纤维素（Carboxymethylhydroxyethyl cellulose）

纤维素（Cellulose）

十六烷基三甲基溴化铵（Cetyltrimethyl ammonium bromide）

聚氨基葡糖（Chitosan）

胆碱（Choline）

柠檬酸（Citric acid）

$N,N-$己二烯$-N-$烷基$-N-$（磺基烷基）铵甜菜碱（$N,N-$Diallyl$-N-$alkyl$-N-$(sulfoalkyl) ammonium betaine）

2,2-二溴-3-次氮基丙酰胺（2,2-Dibromo-3-nitrilopropionamide）

乙二胺四亚甲基膦酸（Diethylenetriamine tetramethylene phosphonic acid）

二羟基异亚氨基$-N,N-$二乙酸（Dihydroxyisopropylimino$-N,N-$diacetic acid）

二巯基噻二唑（2,5-Dimercapto-1,3,4-thiadiazole）

$N,N-$二甲基丙烯酰胺（$N,N-$Dimethylacrylamide）

二甲氨基甲基丙烯酸乙酯（Dimethyl aminoethyl methacrylate）

$N,N-$二甲氨基丙烷基甲基丙烯酰胺（$N,N-$Dimethyl aminopropyl methacrylamide）

二烯丙基二甲基氯化铵（Dimethyl diallyl ammonium chloride）

4,4-二甲基噁唑烷（4,4-Dimethyl oxazolidine）

氯化$N-$十二烯基$-N,N-$二(2-羟乙基)$-N$甲基氯化铵（$N-$Dodecene$-1-$yl$-N,N-$bis(2-hydroxyethyl)$-N-$methylammonium chloride）

十二烷基苯磺酸（Dodecyl benzene sulfonic acid）

十二烷基三乙基溴化铵（Dodecyltriethylammonium bromide）

己六醇（Dulcitol）

二十二烯氨基丙烷基甜菜碱（Erucyl amidopropyl betaine）

$N-$二十二烯$-N,N-$双(2-羟乙基)$-N-$甲基氯化铵（$N-$Erucyl$-N,N-$bis(2-hydroxyethyl)$-N$methyl ammonium chloride）

赤藓糖醇（Erythritol）

乙氧基壬基苯酚(Ethoxylated nonyl phenol)
乙二胺四乙酸(Ethylenediamine tetraacetic acid)
2-乙基己基丙烯酸酯(2-Ethylhexyl acrylate)
甲酸(Formic acid)
富马酸(Fumaric acid)
糠醇(Furfuryl alcohol)
葡萄糖酸(Gluconic acid)
葡萄糖酸-δ-内酯(Glucono-δ-lactone)
D-葡萄糖(D-Glucose)
D-葡萄糖醛酸(D-Glucuronic acid)
油酸(Glyceric acid)
甘油(Glycerol)
N-甘油亚氨基-N,N-二乙酸(N-Glycerylimino-N,N-diacetic acid)
乙醇酸(Glycolic acid)
乙二醛(Glyoxal)
瓜尔胶(Guar)
十六烷基溴化铵(Hexadecyl bromide)
羟基乙酸(Hydroxyacetic acid)
羟乙基乙二胺四乙酸(Hydroxyethylenediamine tetraacetic acid)
羟乙基乙二胺三乙酸(Hydroxyethylethylenediaminetriacetic acid)
1-羟基亚乙基-1,1-二磷酸(1-Hydroxyethylidene-1,1-diphosphonic acid)
羟乙基-三羟丙基乙二胺(Hydroxyethyl-tris-(hydroxypropyl)ethylenediamine)
N-(3-羟丙基)亚氨基-N,N-二乙酸(N-(3-Hydroxypropyl)imino-N,N-diacetic acid)
N-(2-羟丙基)亚氨基-N,N-二乙酸(N-(2-Hydroxypropyl)imino-N,N-diacetic acid)
2-咪唑啉硫酮(2-Imidazolidinethione)
环己六醇(Inositol)
异噻唑啉(Isothiazolin)
4-巯基噻唑烷-2-酮(4-Ketothiazolidine-2-thiol)
乳酸甘(Lactic acid)
十二烷基甜菜碱(Lauryl betaine)
氟化镁(Magnesium fluoride)
过氧化镁(Magnesium peroxide)
顺丁烯二酸(Maleic acid)
羟基丁酸(Malic acid)
丙二酸(Malonic acid)
苯基乙醇酸(Mandelic acid)
甘露糖醇(Mannitol)
α-D-(+)-甘露糖(α-D-(+)-Mannose)
巯基苯并咪唑(Mercaptobenzimidazole)
2-巯基苯并咪唑(2-Mercaptobenzoimidazole)
2-巯基苯并噻唑(2-Mercaptobenzothiazole)
2-巯基苯并噁唑(2-Mercaptobenzoxazole)
2-巯基噻唑啉(2-Mercaptothiazoline)
2-甲氧基乙基亚氨基-N,N-二乙酸(2-Methoxyethylimino-N,N-diacetic acid)
3-甲氧基丙基亚氨基-N,N-二乙酸(3-Methoxypropylimino-N,N-diacetic acid)
甲基亚氨基-N,N-二乙酸(Methylimino-N,N-diacetic acid)
2-甲基油酸(2-Methyl oleic acid)
次氨基三乙酸(Nitrilo-triacetic acid)
十八烷基三氯甲硅烷(Octadecyltrichlorosilane)
乙二酸(Oxalic acid)
P,P'-双苯磺酰肼(P,P'-Oxybis(benzenesulfonyl hydrazide))
邻苯二甲酰亚胺(Phthalimide)
聚(二烯丙基二甲基氯化铵)(Poly(dimethyl diallyl ammonium chloride))
2-甲基油酸钾(Potassium 2-methyl oleate)
过硫酸钾(Potassium persulfate)
吡啶-N-氧化物-2-酮(Pyridine-N-oxide-2-thiol)
N-巯基吡啶-2-硫醇(N-Pyridineoxide-2-thiol)
次氯酸钠(Sodium hypochlorite)

过硫酸钠(Sodium persulfate)
水杨酸钠(Sodium salicylate)
去水山梨糖醇单油酸酯(Sorbitan monooleate)
山梨糖醇(Sorbitol)
丁二酸(Succinic acid)
琥珀酰聚糖(Succinoglycan)
氨基磺酸(Sulfamic acid)
牛脂酰氨基丙基胺(Tallowamidopropylamine)
酒石酸(Tartaric acid)
四羟甲基硫酸磷(Tetrakishydromethyl phosphonium sulfate)
四-N-丙基锆(Tetra-n-propyl zirconate)
1,3,4-噻二唑-2,5-二硫醇(1,3,4-Thiadiazole-2,5-dithiol)
2-硫代咪唑啉酮(2-Thioimidazolidone)
对甲苯磺酰肼(p-Toluene sulfonyl hydrazide)
邻甲基苯甲酸(o-Tolulic acid)
原乙酸三甲酯(Trimethyl orthoacetate)
三(羟甲基)甲基亚氨基-N,N-二乙酸(Tris(hydroxymethyl)methylimino-N,Ndiacetic Acid)
N-乙烯内酰胺(N-Vinyl lactam)
乙烯膦酸(Vinylphosphonic acid)
N-乙烯吡咯烷酮(N-Vinyl-2-pyrrolidone)
乙烯磺酸(Vinyl sulfonic acid)

术语

吸附(Absorption)
表面活性剂(Surface active agent)
水吸附(Water Absorption)
酸压裂(Acid fracturing)
酸化(Acidizing)
胶黏包覆材料(Adhesive-coated material)
支撑剂(Proppant)
吸附作用(Adsorption)
聚合物(Polymer)
沥青质(Asphaltenes)
稳定性(Stabilization)
细菌(Bacteria)
杀菌剂(Biocides)
化学处理(Chemical treatment)
可控降解(Controlled degradation by)
检测(Detection)
环境适应性(Environmental adaptation)
烃氧化(Hydrocarbon-oxidizing)
假单胞杆菌(*Pseudomonas elodea*)
固着(Sessile)
硫酸盐还原(Sulfate-reducing)
硫酸盐还原(Sulfidogenic)
野油菜黄单胞菌(*Xanthomonas campestris*)
细菌控制(Bacteria control)
杀菌剂(Bactericides)
剂量(Doses)
硫酸盐还原菌(For sulfate-reducing bacteria)
压裂液(Fracturing fluids)
选择(Selection of)
可生物降解的配方(Biodegradable formulations)
巴克利—莱弗里特方程(Buckley-Leverett equations)
缓冲(Buffer)
陶粒(Ceramic particles)
螯合(Chelate)
螯合剂(Chelating agents)
可生物降解的(Biodegradable)
干扰(Interference)
耐化学性(Chemical resistance)
色谱法(Chromatography)
体积排除(Size exclusion)
黏土稳定(Clay stabilization)
涂层(Coating)
黏土稳定(Clay stabilization)
胶囊(Encapsulation)

支撑剂(Proppants)
连续油管(Coiled tubing)
缩合产物(Condensation products)
羟基乙酸(Hydroxyacetic acid)
可控水溶性化合物(Controlled solubility compounds)
交联剂(Crosslinking agents)
硼酸(Boric acid)
多价金属阳离子(Polyvalent metal cations)
清除岩屑能力(Cuttings removal)
循环伏安法(Cyclic voltammetry)
消泡(Defoaming)
破乳剂(Demulsifier)
有效性(Effectiveness)
扩散(Diffusion)
气体(Gas)
孔隙压力(Pore pressure)
页岩稳定(Stability of shales)
置换(Displacement)
乳化稳定剂(Emulsion stabilizers)
岩石(Rock)
两相非混相(Two-phase immiscible)
白云岩(Dolomite)
动态张力梯度(Dynamic tension gradient)
乳化剂(Emulsifier)
环氧树脂(Epoxide resin)
刻蚀(Etching)
发酵(Fermentation)
有氧(Aerobic)
滤失速度(Filtration rate)
幂律关系(Power law relationship)
流速(Flow rate)
液体滤失(Fluid loss)
瓜尔胶(Guars)
机理(Mechanism)
液体滤失添加剂(Fluid loss additive)
可控制降解(Controlled degradable)
降解(Degradation)
酶降解(Enzymatically degradable)
吉兰(Gellan)

木质素磺酸盐(Lignosulfonate)
聚(羟基乙酸)(Poly(hydroxyacetic acid))
淀粉(Starch)
单宁酚醛树脂(Tannic-phenolic resin)
泡沫(Foam)
膨胀弹性(Dilatational elasticity)
液膜变薄(Film thinning)
压裂作业(Fracturing jobs)
马仑高尼效应(Marangoni-effect)
稳定性(Stability)
泡沫稳定性(Foam stability)
地层伤害(Formation damage)
气井(Gas wells)
裂缝导流能力(Fracture conductivity)
压裂(Fracturing)
煤层(Coal-beds)
效率(Efficiency)
液体滤失(Fluid loss)
铁离子控制(Iron control)
摩阻损失(Friction loss)
石榴石型(Garnet type)
破胶剂(Gel breaker)
酸(Acid)
络合剂(Complexing agents)
氧化(Oxidative)
乙酸钠(Sodium acetate)
稠化剂(Gelling agent)
颗粒剂(Granules)
砾石充填(Gravel packing)
浆液(Grouting)
水平井(Horizontal well)
界面张力(Interfacial tension)
动力学(Kinetics)
交联(Of crosslinking)
降解(Of degradation)
膨胀(Of swelling)
数学模型(Mathematical models)
代谢(Metabolism)
胶束(Micelle)
微生物种群(Microflora)

微生物(Microorganisms)
防止(Combating)
腐蚀(Corrosion)
矿物油(Mineral oil)
矿化度(Mineralization)
混合金属氢氧化物(Mixed metal hydroxides)
水平井(Horizontal wells)
制备(Preparation)
热能激活(Thermally activated)
钻井液(Muds)
可生物降解(Biodegradable)
混合金属氢氧化物(Mixed metal hydroxides)
页岩抑制(Shale inhibition)
营养(Nutrients)

模拟(Simulation)
水溶性(Water-soluble)
渗透性(Osmosis)
渗透膨胀(Osmotic swelling)
质子化反应(Protonation reaction)
米糠提取物(Rice bran extract)
阻垢剂(Scale inhibitor)
胶囊(Encapsulated)
剪切黏度(Shear viscosity)
污泥(Sludge)
表面活性剂(Surface active agents)
表面张力(Surface tension)
稠化剂(Thickener)
蜡状颗粒剂(Wax granules)

国外油气勘探开发新进展丛书（一）

书号：3592
定价：56.00元

书号：3663
定价：120.00元

书号：3700
定价：110.00元

书号：3718
定价：145.00元

书号：3722
定价：90.00元

国外油气勘探开发新进展丛书（二）

书号：4217
定价：96.00元

书号：4226
定价：60.00元

书号：4352
定价：32.00元

书号：4334
定价：115.00元

书号：4297
定价：28.00元

国外油气勘探开发新进展丛书（三）

书号：4539
定价：120.00元

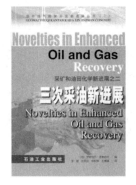

书号：4725
定价：88.00元

书号：4707
定价：60.00元

书号：4681
定价：48.00元

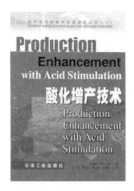

书号：4689
定价：50.00元

书号：4764
定价：78.00元

国外油气勘探开发新进展丛书（四）

书号：5554
定价：78.00 元

书号：5429
定价：35.00 元

书号：5599
定价：98.00 元

书号：5702
定价：120.00 元

书号：5676
定价：48.00 元

书号：5750
定价：68.00 元

国外油气勘探开发新进展丛书（五）

书号：6449
定价：52.00 元

书号：5929
定价：70.00 元

书号：6471
定价：128.00 元

书号：6402
定价：96.00 元

书号：6309
定价：185.00 元

书号：6718
定价：150.00 元

国外油气勘探开发新进展丛书（六）

书号：7055
定价：290.00 元

书号：7000
定价：50.00 元

书号：7035
定价：32.00 元

书号：7075
定价：128.00 元

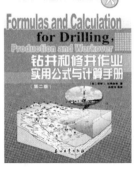

书号：6966
定价：42.00 元

书号：6967
定价：32.00 元

国外油气勘探开发新进展丛书（七）

书号：7533
定价：65.00元

书号：7802
定价：110.00元

书号：7555
定价：60.00元

书号：7290
定价：98.00元

书号：7088
定价：120.00元

书号：7690
定价：93.00元

国外油气勘探开发新进展丛书（八）

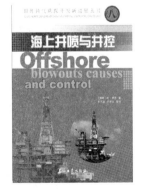

书号：7446
定价：38.00元

书号：8065
定价：98.00元

书号：8356
定价：98.00元

书号：8092
定价：38.00元

书号：8804
定价：38.00元

国外油气勘探开发新进展丛书（九）

书号：8351
定价：68.00元

书号：8782
定价：180.00元

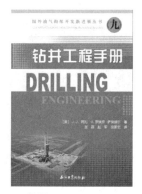

书号：8336
定价：80.00元

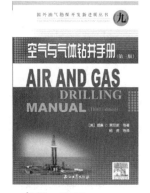

书号：8899
定价：150.00元

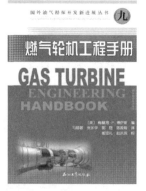

书号：9013
定价：160.00元

书号：7634
定价：65.00元

国外油气勘探开发新进展丛书（十）

书号：9009
定价：110.00

书号：9989
定价：110.00

书号：9574
定价：80.00

书号：9024
定价：96.00

书号：9322
定价：96.00

书号：9576
定价：96.00

国外油气勘探开发新进展丛书（十一）

书号：0042
定价：120.00

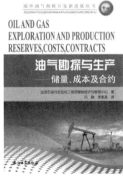

书号：9943
定价：75.00

书号：0732
定价：75.00

国外油气勘探开发新进展丛书（十二）

书号：0916
定价：80.00

书号：0916
定价：80.00

书号：
定价：

国外油气勘探开发新进展丛书（十二）

书号：0661
定价：80.00